Getting Started with
DAVINCI RESOLVE 18

Basic Video Production for YouTube and Social Media

Henry J. James

ANODYNE
P R E S S

Copyright © MMXXII Anodyne Press, LLC
All rights reserved

Notice of Rights
All rights reserved. No part of this book may be reproduced or transmitted in any form by any means, mechanical, electronic, photocopying, or otherwise without prior written permission of the publisher.

Notice of Liability
Neither the author nor the publisher shall be liable for any loss or damage caused (or alleged to be caused) by information contained herein or on any websites or other media controlled by the publisher or author.

Intellectual Property
Where the publisher was aware of a trademark claim, those are marked as requested by the owner of the trademark. All other product names and services identified in this book are used in editorial fashion only and for the benefit of such companies with no intention of infringement of the trademark. No such use, or the use of any trade name, mark, device, or logo, is intended to convey endorsement of or other affiliation with this book.

DaVinci Resolve and the DaVinci Resolve logo are registered trademarks of Blackmagic Design Pty Ltd.

Adobe and Adobe Premier are either registered trademarks or trademarks of Adobe, Inc.

Apple, the Apple logo, iMac, MacBook Pro, macOS, Mac, and Final Cut are either registered trademarks, trademarks, or service marks of Apple, Inc.

Windows is a registered trademark of Microsoft, Inc.

Pro Tools is a registered trademark of Avid Technology, Inc.

Credits: Film Editor Brad Mays at the Moviola—Creative Commons 3.0.

All other photos Shutterstock or author submitted.

Layout and design by Adam Robinson at Good Book Developers

ISBN 978-1-945028-46-5

Printing 9 8 7 6 5 4 3 2 1
Version 1.00

For updates, errata, or business inquiries, visit www.anodynepress.com

ANODYNE
PRESS

Dedication

The fellows who taught me television production all hailed from New York City. They had been active there during television's Golden Age—the time of *Playhouse 90*, *Marty*, *Requiem for a Heavyweight*, and *No Time for Sergeants*–the era when Paddy Chayefsky, Rod Serling, and John Frankenheimer ruled. When these guys were doing TV, it was live—sometimes with as many as 10 cameras—and all in glorious black and white. Often, they did two shows back-to-back—an early live show for the east coast and another live show three hours later for the west coast—because videotape hadn't been invented yet.

But in 1956, Ampex developed the first practical videotape recorder, and that changed everything—including where TV was made. Production started moving west—to Television City in Hollywood and NBC Burbank. But this tight-knit group of directors and editors stayed in New York. (Once, they did travel to Moscow to record the famous "Kitchen Debate" between Vice President Richard Nixon and Nikita Khrushchev on Ampex *color* videotape—a first.)

Of course, by the time I met them, they were semi-retired—but still very much in the game. I learned a lot about TV production from them, like how to edit two-inch quad tape with a (degaussed) razor blade so that every TV set in America wouldn't jump when the splice hit the head wheel spinning at 14,400 rpm—and sometimes that actually worked. I learned how to shoot and edit film, how to direct live TV, and eventually how to edit videotape *electronically* (using reel-to-reel machines).

The man who taught me TV lighting (or at least took a stab at it), Imero Fiorentino, was a genius. Everybody called him "Immie," although, ironically, he never won one. He should have. He's the guy who perfected Chroma Key. Without his lighting techniques, it just didn't work. When something difficult needed to be lit, he didn't rush off to the rental house. He went to the hardware store. Fiorentino's main principle was that lighting was what it looked like on camera, not what it took to get there. That's a lesson I've never forgotten. His bio, *Let There Be Light* is worth reading if you care anything at all about television's Golden Age.

It's to Murray, Clyde, Bill, Immie, and the rest that I dedicate this book.

—HENRY JAMES

Contents

Getting Started with DaVinci Resolve 18	1
Shooting for DaVinci Resolve	19
Recording Audio for Resolve	55
DaVinci Resolve Setup	65
Exercises	93
Exercise Uno	108
Exercise Due	135
Exercise Tre	149
Exercise Quattro	165
Pages	179
The Media Page	181
The Cut Page	188
The Edit Page	196
The Fusion Page	210
The Color Page	220
Fairlight	228
The Deliver Page	249
Backups and Archiving	263
Resources	271
Hardware Requirements & Configuration	279
Resolving Performance Issues	297
Render Systems	313

Getting Started with DaVinci Resolve 18

Getting Started with DaVinci Resolve

DaVinci Resolve 18 from Blackmagic Design is the post-production system of choice if you are serious about doing video. There are simpler editing systems, but IMO the rest don't hold a candle to Resolve (and there certainly aren't any cheaper ones). I believe that once you master any of those less-capable video editors, you'll be discouraged by their limitations. Better, in my view, to be frustrated by complexity than austerity.

And DaVinci Resolve has plenty of complexity. Resolve started life as a color grading system for film transfer and restoration (da Vinci). Blackmagic Design (BMD) beefed up the basic editing tools, added Fusion (effects), Fairlight (audio), and some other bells and whistles to create a complete all-in-one video post-production system.

If you are a current user of Adobe Premier Pro or Final Cut Pro, I think you are in for a pleasant surprise. Yes, for a while, at least, muscle memory will not be your friend (although you can set the keyboard in Resolve to mimic Premier or Final Cut). But the basic editing functions will be familiar to you, and DaVinci color grading and Fusion effects are game-changers.

What's New With DaVinci Resolve 18?

Not that much, actually (and that's a good thing). With this new version, DaVinci Resolve has now reached a level of maturity such that Resolve 18 is almost indistinguishable from Resolve 17—*on the surface*. The big change is what happened under the hood.

By Resolve 16 and 17, BMD had added so many features that Resolve needed to be run on a fairly powerful computer, or it stuttered and glitched during playback. Renders were frequent and often took a long time. Having powerful hardware made a difference. Preference settings were important, and many users had to operate in restrictive ways to avoid their computer bogging down during playback or waiting until the edit was nearly finished before adding complex effects.

However, with Resolve 18, BMD has done a lot of work inside Resolve, and it now runs much faster (with fewer, faster renders) than previous versions, even on less powerful hardware. This means that the hardware requirements are far less, and the setup/installation of Resolve on your computer is less critical than with previous versions.

This is important because digital video editing systems spend a lot of time rendering. (There's a whole chapter on Resolve's various rendering systems, to help you learn more about them.) In a nutshell, rendering means Resolve has to convert your camera's compressed video into honest-to-god frames in order to let you edit it, apply effects, or even just play it back. Some camera compression systems (called "codecs") are harder to playback than others. Resolution is also a factor. A 4K clip has a lot more information that has to be processed than 1080p. In many cases, Resolve can often play video from your camera on the fly without having to pause and process each frame (depending on your computer's CPU, GPU and disk drive speed, your camera's recording/compression format/codec, the video's resolution and frame rate, and various Resolve settings such as Timeline resolution.)

If Resolve needs to render a video clip before it will play without fits and starts, a red line will appear over the clip. As Resolve renders the clip, the line will turn blue. There are various settings that enable or disable this behavior—see the Rendering chapter for details.) So the rewrite of the underlying code that performs the renders is a big deal.

Resolve 18 will run on most PCs and Macs with just the default settings and without you having to know very much about Resolve's various rendering systems. In previous versions of this book, there was a lot more information upfront about hardware requirements, various rendering systems, and resolving performance issues. Some users will still experience performance problems and will need to make sure Resolve 18 is set up properly on their computer. Some users will need better hardware. And some will still experience performance (playback) issues that will need to be addressed. But most won't. It all depends on how your camera records video (codec), how powerful your computer is, and how complex the effects you add are. For this reason, the hardware requirements, render system operations, and performance tuning have been moved to the end of the book. They are there if you need them, but with the rewrite of the Resolve 18 render engine, my guess is that won't be many of you.

In addition to the improved render system, a few new features have been added to Resolve 18, such as "Cloud Collaboration." Cloud Collaboration could be handy if you are working on projects with multiple editors located around the globe. That's not me, and I'm guessing, if you are a new user, that's not you either. However, BMD has also added several AI components to existing effects that make them so easy to use they are almost magical. I will nevertheless cover how to manually use these effects (such as Tracking), but the new AI-powered *Intuitive Object Mask* is mindboggling.

That said, the UI for Resolve 18 is extremely similar to 16 and 17, if not identical. However, BMD makes minor changes very frequently (and occasionally major ones as well), so buttons and tabs tend to get moved around from time to time, and names get changed.

Pages

Resolve consists of seven functional areas BMD calls "pages." The main pages are Media, Cut, Edit, Fusion, Color, Fairlight, and Deliver.

The editing process begins (left to right) on the **Media** page, which is used to import video, stills, audio tracks, and other media (although media can be imported via other pages, too).

The **Cut** page has everything you need to edit a video all on a single page. The Cut page was developed for TV news departments who need a limited number of things done very quickly and repetitively. BMD touts the Cut page for beginners, but I don't recommend Cut page for novices or even general use for reasons I explain elsewhere. (See the Cut page chapter for more info.)

The focus of the **Edit** page is editing, obviously. But throughout Resolve, you'll see that you can often perform other functions, sometimes with limitations, sometimes not, on just about any page, regardless of the name.

World-class video effects and animation are handled through the **Fusion** page (but basic effects and even some Fusion effects have now been added to the Edit and Color pages).

The **Color** page is primarily for grading, color correction, and OpenFX effects. My guess is you'll spend most of your time on the Edit and Color pages.

Fairlight is a complete digital audio workstation (DAW). However, there are ample audio tools on the Edit page, and it's unlikely you'll need Fairlight unless you need to make use of its 2,000-track capability. With Resolve 17, Pro Tools shortcuts were added (which, for some of you, at least will make Fairlight easier to learn).

The **Deliver** page has the tools for compressing/encoding the finished product and for uploading it to YouTube, Vimeo, Twitter, or other social media platforms (or for simply making an output file).

Although not organized as "pages," Resolve also has excellent tools for backing up and archiving media or entire projects.

Versions

There are two main versions of DaVinci Resolve (which you can download from www.blackmagicdesign.com). If you are just starting out, I'd suggest getting the free version and test driving it. The free version is wonderful, but a few features like an always-on full-screen monitor and the ability to use more than one graphics card are absent (and a few other things, but very few, actually). And some effects are watermarked.

Many users report that the underlying code in Studio is faster than the free version. There are probably other improvements under the hood as well. As John's Films, a YouTube channel that focuses on the hardware aspects of Resolve, puts it, it's probably cheaper to upgrade to Studio for a performance boost than buying better hardware. His tests showed that there was no appreciable difference in render speeds for a $300 graphics card and one that cost more than a thousand, but Studio was about 30% faster than the free version at rendering on either card. That said, Resolve 18's inner workings were completely rewritten, and both the free and paid versions of Resolve 18 run much better (fewer glitches and shorter renders) on less than state-of-the-art hardware than previous editions.

I bought Studio because I wanted a full-time, full-screen monitor, and with the free version, you'd need a DeckLink card ($150) to do that. So Studio cost me only $150 net when you consider I didn't have to buy a DeckLink card. And when you buy Resolve, it's yours forever (for less than the cost of a year's worth of Adobe Premier Pro CC). And you get free (and frequent) updates. BMD usually gives away Studio for free if you buy their Speed Editor controller. So for $300, you can often get a really nice edit controller plus Studio.

Learning Resolve

If you are just learning to shoot and edit video, Resolve will be a challenge, but not an overwhelming one. In my view, you'd have nearly as much trouble starting with Adobe Premier Pro or Final Cut Pro as you're going to have with Resolve.

Because of its capabilities and features, some people think Resolve is intended only for professionals. Well, it is, and it isn't. Yes, it's a professional video production system with loads of high-end features. But even so, the basics of Resolve are fairly easy to learn. And learning Resolve will pay dividends as your skills (and gear) improve. Once you've mastered the fundamentals, Resolve will be able to handle anything you are ever likely to want to do with video and audio. And ironically, if you have less-than-stellar gear, Resolve can let you get the absolute top performance out of your equipment and then some.

One source of frustration when approaching Resolve for the first time is that there's a lot of outdated information online. If you research Resolve on Google or YouTube, you'll find that many of the complaints and fixes refer to issues with earlier versions and problems that no longer exist. While there's a wealth of information on YouTube and blogs, make sure you know which version the advice purports to cover. (See the Resources chapter for further information.)

The basic operations of Resolve 16, 17, and 18 are very similar, and most users will have no problem moving from version to version. The biggest changes to DaVinci Resolve really started with Resolve 15. Later editions cleaned things up a bit, added a few new features, and of course, moved things around. But for basic operations, DaVinci Resolve 16, 17, and 18 are nearly identical.

BMD, unlike many of its competitors, continually makes improvements and adds many new features with every release. BMD updates Resolve frequently, so depending on which version of Resolve 18 you have, the UI could be slightly different. The constant updates are a source of frustration for some but a joy to many. And if BMD keeps up the pace, Resolve will only get better and better year after year. Compare with Adobe Premier Pro (where the only real changes have been price increases) or Final Cut Pro (once my favorite editing system), which Apple completely, intentionally, and totally f***ked up. But I digress.

FREE BOOKS!

Blackmagic Design also publishes a *DaVinci Resolve Reference Manual* (in PDF). It's nearly 4,000 pages, but it's free. The Reference Manual is comprehensive and was written for professional users, but it can be a godsend if you're stuck. (You can also get to it from the Help tab.)

There's also a book called *Beginner's Guide to Davinci Resolve* (500 pages) that is also available as a free download (in PDF). In fact, AFAIK, all of BMD's manuals and books are available on their website as PDFs, and you can download them all for free. And if that's not enough, BMD has an online training course with more than 20 well-produced videos—also free.

Updates

DaVinci Resolve combines three programs that were originally stand-alone. The Color page is essentially DaVinci color grading. With version 12, BMD started adding simple editing tools. Those tools improved dramatically with later versions. BMD eventually grafted on the compositing/effects system known as Fusion. Then, beginning with Resolve 16, BMD added Fairlight, a complete Digital Audio Workstation (DAW). The most current release is Resolve 18. But there are many who still use Resolve 16/17 or even earlier versions.

Most of the upgrades in Resolve 16, 17, and 18 are there to make the effects artists at Disney, Marvel, ILM, and DreamWorks happy. And I'm sure they're ecstatic. The new Resolve 18 is loaded with wowie-zowie effects. More importantly, 17, 17.4, and 18 fixed a lot of problems that users complained about with 16 and earlier versions. I won't go into all that—the forums and Reddit cover that *ad nauseam*.

BMD makes frequent updates to Resolve, and the latest version of anything is usually not the most stable and can introduce new problems while trying to fix older ones, so be cautious about accepting updates. If Resolve suddenly stops working the way it used to, my advice is restart Resolve, and if that doesn't fix it, reboot. Often, that's all it takes. And if Resolve does "crash," selecting "Wait for Program to Respond" often gets things back on track. Sometimes, Resolve seems to get lost in deep thought.

About This Book

There are so many things Resolve can do and so many different ways to do them that it was hard to know what to cover and what to skip. Throughout this book, I've had to constantly remind myself that it's an *introduction*. If I had tried to cover everything that anyone could possibly want to do with Resolve, I'd have ended up with a nearly 4,000-page manual like Blackmagic did. Instead, I have intentionally tried to focus on the essentials that will familiarize you with the basics of how Resolve works without overwhelming you at every turn.

However, there is a lot of information in this book that many of you may not need or may not need right now. In that case, feel free to skip it.

Previous versions of Resolve, particularly Resolve 16 and 17, required users to pay careful attention to how Resolve was set up in order for it to play the Timeline smoothly on underpowered PCs (and virtually all Macs). That's much less of a problem with Resolve 18. While I think you probably should know and understand what the system and user preferences do, many users can simply accept the defaults and start editing. In fact, you can probably just jump from here to the first lesson if you want.

However, there are chapters about shooting video and recording audio that you may want to review if you haven't had very much experience creating videos. Many beginning video makers use Resolve to fix problems with their video and audio that could have been handled at the camera or mic level. Of course, all of us have to do that from time to time because of bad lighting or noisy locations, but it's bad practice to spend most of your time and effort using Resolve's powerful tools to fix problems you could have avoided from the get-go.

In the interest of completeness, there are chapters on the Cut page and Fairlight audio, even though I don't think beginners should use them for reasons explained later.

> **CAVEAT**: The information in this book was sourced from more than a dozen professional editors, each of whom has years of experience with Resolve. Every keystroke and mouse click was tested on Resolve 16, 17, and 18 (Studio version). Obviously, some features and functions may not work at all in the free version or may work differently (usually slower). There are the usual differences between the Mac and Windows versions, too (mostly involving shortcut keys, the User Interface, and ProRes). But keep in mind that BMD makes frequent changes to Resolve, and with each new release and update, you might have to re-interpret some of the information herein, depending on what build you have. If something doesn't work as described, visit **anodynepress.com** for possible explanations and solutions as well as updates.

Adobe Premier Pro and Final Cut Users

Unlike Adobe Premier Pro, Resolve has everything under one roof. It combines a lot of the functionality of After Effects, Audition, and even Pro Tools in a single, integrated app.

If you are already quite familiar with Premier Pro, Final Cut Pro, or other comparable Non-Linear-Editors (NLEs), you'll probably think this book is for beginners. It is. In one sense, it's for people who have never used any video editing system except perhaps iMovie or Windows Movie Maker and want more than those consumer editors can deliver. But it's also a very straightforward introduction to Resolve without overwhelming you with overly detailed descriptions about everything Resolve can do. And if you've been using Resolve for a while and still can't seem to find the controls necessary to do what you want, this book will also help you.

YouTubers and WordPress Vloggers

If you are creating videos for YouTube, Vimeo, Twitter, or WordPress, the exercises in the book will be directly applicable. At the end of the day, you need to be able to import media from your camera into Resolve, edit the video, apply any effects you want, and output a finished file that will upload to YouTube or wherever you want to show it. For that reason, the lessons are "end-to-end" rather than simply focusing on what happens in the middle.

In **EXERCISE UNO**, you'll shoot and edit a straightforward "talking head" video, typical of what you might find on YouTube. Nothing could be simpler. But in doing so, you'll learn how to import media, how the Media, Edit, Color, and Deliver pages work, creating and copying Nodes, basic color correction, navigating the Timeline, using text and titles, setting audio levels and EQ, as well as encoding for YouTube, Vimeo, Twitter or wherever. In just this one exercise, you'll step through the entire gamut of video production from end to end because you need to know *all these things,* even if you are just making a straightforward YouTube video.

For **EXERCISE DUE**, you'll shoot a Chroma Key video (green screen) and composite it with a background plate. You'll learn the basics of grading and creating garbage mattes and compound clips. Plus, you'll reuse everything you learned in the first lesson.

EXERCISE TRE teaches common editing tools and techniques as well as effects often used in YouTube videos and by WordPress bloggers. Specifically, this exercise will give you some experience grading footage and improving the look of your videos with effects such as Beauty and Face Refinement using both Tracker and Qualifier as well as Intuitive Object Mask, which is new for Resolve 18. Once you know how to operate Power Windows, Qualifier, and Tracker, you'll have no trouble applying hundreds of effects.

EXERCISE QUATTRO covers object tracking, blurring, and a few other tools like Object Removal and Patch Replacer. There are also exercises using smartphone video, making and using stills, as well as creating thumbnails for YouTube.

What Effect?

I should mention that when you apply an effect to a clip, it's usually applied to the entire frame. With some effects, that's what you want. But many effects are intended to be applied to only a portion of the frame, such as a face or some object in the frame. In that case, before applying the effect, you have to tell Resolve *where* to apply it (and where not to). That's done with various "qualifiers." Chroma key (green screen), for example, uses a bright color (green) to "qualify" where the background image should go (and not go). Another qualifier is Window. Here you decide what goes where based on common shapes (circle, square) or a freehand outline drawing. One of the most powerful qualifiers is Resolve's ability to track an object (face, license plate, basketball—whatever) and apply an effect. In this case, you use a qualifier to identify the object (drawing a line around it, for example) and then use Tracker to follow it as it moves. With Resolve 18, BMD introduced *Intuitive Object Mask* that automates this often problematic process and achieves astounding results. We'll cover both methods.

If you can shoot a simple video, import it into Resolve, perform basic edits, perhaps add an effect or two and output a file that will upload to YouTube or similar, you are really halfway there. From that moment on, the rest will make sense to you, and from that solid base, you'll find it easy to add to your expertise by watching effects tutorials or reading BMD's official guides. That, in a nutshell, is the idea behind this book.

While the exercises in this book are basic, they cover a lot of ground. It is expected that you'll *repeat each of the exercises* three or four times until you can do them quickly and efficiently—without even thinking about it—because *repetition is key to learning*. If you are experienced using Premier Pro or Final Cut Pro, you'll only have to work through each exercise once. Or maybe a second time just to make sure you've got it. OK, third time's the charm.

I've heard new users complain about Resolve and its supposed "learning curve." Yes, Resolve has a learning curve, and true, it is probably steeper than most. But it's a hill, not a mountain. If you set as your goal creating a simple video from end to end, follow the step-by-step exercises in this book, and *repeat them several times* (you know, *practice*), you'll quickly (and painlessly) master Resolve. The underlying principle of this book is that practicing a few rather simple lessons over and over again will let you master Resolve in ways that trying 30 different things only once will not. I know from experience that doesn't work. And the learning curve is not just a Resolve issue. If you've never used a *professional* video editing system before, I assure you that learning Premier, Final Cut, Avid, or one of the others, is no picnic either.

There are hundreds of YouTube videos with tutorials on everything from motion graphics to creating a cinema look. But almost all of them assume you have at least a minimal knowledge of Resolve, and they can be hard to follow if you don't. But once you know how to do even one complete end-to-end project using the tools on the Media, Edit, Color, and Deliver pages, you'll be able to apply the more advanced techniques from the many excellent Resolve tutorials on YouTube. (See the Resources chapter for a curated list.)

Back to the Future

The first film editing I ever did was on a contraption called a Moviola (you'll have to Google it). All the great films of classic Hollywood, from Gone with the Wind to Citizen Kane, were edited on a Moviola. Even Spielberg edited a film on a Moviola—once. How? I don't know. It's a very cumbersome device. Be glad you've never had to use one.

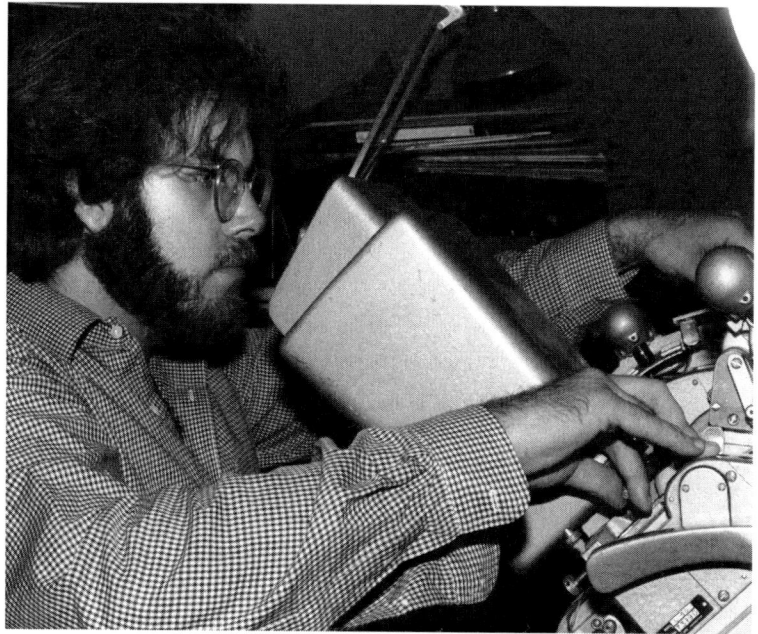

Filmmaker Brad Mays editing his first feature film 'Stage Fright' on an upright Moviola, circa 1987

I never really liked the process of physically editing film. It's too fiddly. It took days to break down the footage onto separate rolls. The audio tracks had to be rerecorded on rolls of sprocketed 35mm magnetic film stock called "full coat." You marked effects and transitions directly on the workprint with a grease pencil. Since the picture and sound were on separate rolls, you had to constantly re-sync everything. And then you had to send the film to a lab and wait two weeks to see the results (which is why I sometimes scoff when younger colleagues complain to me about render times). Plus, I had to use that damned Moviola. (Fast forward and today, moviola.com is a fantastic source for online cinematography courses and filmmaking techniques.)

Later, I got a better-paying job editing film, but only if I could use a flatbed editor. I had never used a flatbed before, but editing is in the mind, not the hand, so I said I could and got the job. During the move to the new job, I stopped off at a post-production house with a couple of rolls of film in my pocket and rented a flatbed edit suite for a couple of hours. Piece of cake.

I started the new job, but within a few weeks, the bosses decided that video was the future, and they were going to move away from film and start doing everything on videotape. Fortunately, I had done a lot of work in video and preferred video over film anyway, so I was all set. It turned out that I never did edit anything on that flatbed editor after all. And a few years later, when I became production manager, I sold the damn thing (along with the unused Moviola).

Over the years, I have had to learn how to use many different video editing systems–Ampex, CMX, Avid, Vegas, Final Cut, Adobe Premier, and others I can't even remember the names of now. So I know firsthand that learning a new editing system can be daunting. It seems as soon as you master one system, a new one comes along.

However, in the case of DaVinci Resolve, I believe we are seeing the editing/post-production system of the future. With thousand-page manuals and hundreds of tutorial videos, there's a lot to learn. But Resolve gets better with every release and will likely be the video editing standard for the next generation. So it will not be a waste of effort to learn Resolve because you will probably be using it 10 years from now—perhaps even longer.

A (Very) Short History of DaVinci Resolve

At one time, da Vinci (that's how they spelled it back in the day) was a room full of custom computer hardware and specialized software used to transfer feature films to tape and DVDs. Da Vinci was mainly used to clean up and restore classic old movies for broadcast and cable. Grant Petty, an Australian *telecine* engineer, operated a da Vinci suite back in the late 1980s. (Telecine is just an old-fashioned word meaning transferring film to videotape.)

The most direct way of transferring film to tape is to simply point a movie projector at a TV camera. And that's basically how a telecine operated. All TV stations had a telecine or "film chain" that converted 35mm slides and 16mm or 35mm film to video.

The aspect ratio of SD (Standard Definition) TV is 4:3 (aka 1.3:1) is really close to the early motion picture standard of 1.375:1, which was adopted when film first learned to talk. By the 1950s, when free TV was simply killing the film industry, studios decided to give the public what TV could not: wider screens and color. Starting around 1953, films were released in Panavision, Cinemascope, VistaVision, and a host of other wide-screen formats shot on Technicolor, Eastmancolor, and other color film stocks.

TV was stuck with 4:3 and B&W, but as early as the late 50s, TV started to colorize a bit (NTSC, PAL, SECAM). However, those first color sets were very expensive, and color television was not commonplace until the early 70s.

Because of the differences in frame rates, aspect ratios, formats, and film stocks, there have always been varying degrees of difficulty transferring film to tape (hence the need for a telecine engineer). Many easily obtainable film prints were not in very good shape (scratches, dirt, faded dyes), and going back to the original negative and restoring a classic film is an extremely expensive process. A studio might do that for *The Wizard of Oz*, *Snow White*, or *On the Waterfront*, but not for *Bwana Devil* or the thousands of other "B" movies and lesser titles.

Because of the costs, many of the movies that ran on TV in the 60s and 70s were just awful—I mean the *prints*—I'm not referring to the content (but in most cases, that opprobrium also applies). Some prints of now-classic or cult films survived only because they were shipped to Alaska for exhibition, and since it was too expensive to ship them back, someone buried them in the frozen tundra. Eventually, they were dug up, shipped back to the lower 48, and scanned on a telecine.

DaVinci (modern spelling) was a color correction system for film transfers. A 35mm film was scanned using lasers and CCDs, and DaVinci hardware and software cleaned up the scan, removed scratches, corrected Gamma, fixed faded color dyes, and out came a pristine videotape at a fraction of the cost of honest-to-god

Hollywood-style film restoration. The master videotape was dubbed to VHS or DVD.

DaVinci was cheaper than archival film restoration, but the gear itself was not cheap. A fully outfitted DaVinci suite could run $800,000 (plus $80,000 a year in maintenance fees). Even 10 years ago, DaVinci Resolve cost a cool $100,000. Today, as you know, it's less than $300, and there is a free version as well.

In the meantime, Petty started Blackmagic Design (BMD) to make high-grade video capture cards for PCs and Macs. The company prospered and eventually expanded to include an entire line of video production and post-production electronics. Today, Blackmagic Design makes some of the best gear in the business.

When DaVinci Systems came up for sale in 2009, BMD bought it and created a version for the Mac and eventually for Linux and Windows. Although BMD remains mostly a hardware company, they have steadfastly worked to improve DaVinci Resolve, such as adding Fusion visual effects.

Later, Fairlight, a complete digital audio workstation (DAW), was included. Originally, Fairlight was a *piano keyboard-based* analog synthesizer built by two Australian teenagers. (It may seem like an odd name for an audio system now, but it was named after the Sydney harbor ferry.)

In attempting to create a digital version with the help of a Motorola engineer, the two teens, Kim Ryrie and Peter Vogel, made digital recordings of actual musical instruments, thus creating the first sampler. By then, Fairlight included software to make it all work. Fairlight was a huge success and was heavily used throughout the recording industry for strings and horns as well as for sampling. The Fairlight package in Resolve is based on the DAW in the original Fairlight CMI system.

DaVinci Resolve, essentially the same program you can buy for $299 (or get for free!), has been used on *Avatar*, *Pirates of the Caribbean*, *Spectre*, *Star Wars-The Last Jedi*, and *X-Men*—all told, more than 300 feature films—several of them Oscar winners or nominees. On the

TV side, DaVinci Resolve is used on *The Big Bang Theory*, *The Man in the High Castle*, *Sons of Anarchy*, *Westworld*, and *The Muppets*. In fact, dozens and dozens of broadcast network, cable, and streaming shows are using DaVinci Resolve Studio right now.

Probably fewer than 100 people were using DaVinci Resolve 10 years ago. Today, it's north of *2 million*. Some of them are Oscar- and Emmy-winning editors. Others are simply guys and gals shooting YouTube videos in their living rooms on a DSLR, camcorder, or smartphone. They all use DaVinci Resolve— the video post-production system for everybody. And so can you.

Shooting for DaVinci Resolve

"Fix it in post." God knows how many times in my life I've heard someone say that (although it eventually became such a cliché, hardly anyone says it anymore except as a joke). Resolve has lots of powerful tools that will let you fix just about anything in post, but you are always better off getting it right *in camera* than trying to fix it later.

Sometimes, due to exigent circumstances, particularly if you are shooting "run and gun" (*kick-bollocks-scramble* for those of you in the UK), you may have to make substantial fixes to the footage when editing. Fine. That happens to all of us at some point. But it's bad practice to take shortcuts when shooting and always having to fix things in post. Fixes take time, almost always introduce unwanted artifacts, and never, ever work as well as video that's been properly shot (or audio that's been properly recorded). The more time you spend fixing things that didn't need to be fixed in the first place, the less time you'll have to work on sequences, storytelling, achieving an esthetic look—whatever your endgame is.

> **CAVEAT:** Several reviewers with extensive experience in video production had concerns about starting off with basic camera setup and video/audio production techniques. Their thinking was that some readers who already have experience with video and audio won't need this information. True enough, and if that's you, you can skim or skip as you like. However, I do see lots of folks use Resolve and other NLEs to fix problems that could (and should) have been avoided in the first place. But it's your book, and you can do what you like.

Workflow Implications

How your footage gets shot, recorded, edited, and distributed is affected by your camera, its settings, Resolve's settings, and your PC's hardware. What follows is designed to help you establish a workflow appropriate for your gear. But it's easy to get lost in the weeds, so before getting further into the details, let me stop here and give you a few plot spoilers.

If your camera only records in H.264 or similar, your PC may have trouble playing and editing that footage without lots of rendering. (This is less of an issue with Resolve 18, but it still happens.) One option is to use an external recorder and record in a different format intended for post-production. Another option would be to transcode the H.264 footage in Resolve to DNxHR (or ProRes on a Mac). Usually, this is done at a reduced resolution (creating "proxies") to reduce some of the processing burden on your PC. When you have finished the project, Resolve will then use your source footage when making the output file for YouTube or where ever you are sending it.

The choices you make regarding the camera's settings, the use of an external recorder, choice of codec, resolution, and frame rate, as well as a multitude of Resolve settings are important considerations for quality, what sort of computer hardware you'll have to have to run Resolve, and how long renders (particularly the last one) will take. There is no one-size-fits-all solution because the answers depend on your camera, its recording method, your PC's capabilities, and what you are trying to accomplish.

As covered in greater detail below and later, if you are simply having trouble playing the Timeline because of render issues, you can transcode to DNxHR (a format designed specifically for editing), you can create reduced resolution proxy files, or you can reduce the resolution of the Timeline and/or video monitor while you edit. Another option is to record in a post-production codec (in camera or via an external recorder). In the latter case, the footage will be of higher quality as well. To add yet another complication, some cameras (and external recorders) can generate *proxies* on the fly. (You import both files into Resolve and link the proxies to the original footage. You edit using the smaller proxy files, and Resolve uses your camera original footage when creating the final output.

So a common workflow would be to shoot at high resolution using a post-production codec (if available), or transcode to a lower resolution proxy for the edit. In every case, you would output for final using the source footage.

These and other techniques were often needed with previous versions of Resolve because most run-of-the-mill PCs and Macs had issues processing certain effects and playing them back in real-time without having to stop and render or without glitching and stuttering. There are still effects that are so complex that playback will stutter and glitch (or stop and render), but with the rewrite of the underlying render engine code in Resolve 18, this may not be a problem for your PC/Mac. If it's not, you will need to know a lot less about what's going on under the hood with Resolve's various rendering systems. In that case, you can skip the parts about how the render systems work (lucky you). However, if you do experience playback/performance issues, there are ways around those (user preferences, workflow modifications) that will help.

The Camera

YouTubers, vloggers, and beginning filmmakers often gripe about their gear. Usually, it's the camera that's not up to scratch, a lens they don't have, or light kit they wish they had. But there are many things you can do in Resolve to make your camera and other gear punch well above its weight.

Of course, Resolve can do wonderful things if your camera has an Arri, Blackmagic, or RED badge on it. But Resolve is even great at improving the look of less-expensive consumer and prosumer camcorders, DSLRs—even smartphones. In fact, one of life's little ironies is that low-end gear can often benefit more from high-end grading and various Resolve tricks than top-of-the-line equipment—particularly if the scene is properly lit.

Camera Specs

Either you are going to shoot with the camera you have, or you are in the market for a new one. If it's the latter, check out several of the YouTube channels you like the look of and see what gear they are using (and not just the camera review channels, either). That's better than trying to pick the perfect camera based on numerical specs that may be misleading or compensated camera reviews (not all of them are, of course).

I've been buying professional broadcast video gear for decades. My take is that camera specs can be misleading—sometimes intentionally so. Important specs often can't be found, and BS specs are hyped to the high heavens.

But I'm used to that. When I was a kid, transistor radios were sold based on how many transistors were in them. A seven transistor radio was better than a five transistor radio, and a nine transistor radio was even better. That was because transistors were very expensive.

Later, transistors got cheap—very cheap. I remember buying a seven transistor radio and opening it up to discover it was a five transistor radio circuit with two dud transistors soldered to the PCB that had no functionality whatsoever. Technically, it was a "seven transistor radio"—but developed by the sales department, not audio engineering.

That sort of thing still happens all the time, particularly with complex gear like video equipment. It irks me to see a camera listed as 10-bit video but, when you read the fine print, it really only does 8-bits in 4K and, while it is 10-bits in HD, it's actually 8-bit video in a 10-bit wrapper—two dud bits. So what is it really?

My advice about cameras is to figure out what you intend to shoot a lot of (green screen, interiors, weddings, exteriors, sports, travel videos) and get some actual raw footage from those cameras, bring the footage into Resolve, and see how it looks to you.

That's not going to be an easy task since so much test footage has been doctored regardless of claims otherwise or is distributed in file formats different from the camera's. But that's the best way to cut through the marketing BS and opinions and figure out what camera is right for you, your workload, and your process in order to achieve the results you want.

That said, a 4K camera coupled to an external recorder such as the Atomos Ninja V that lets you record log 10-bit 4:2:2 video using a ProRes codec would be ideal for just about any purpose. And Resolve can edit ProRes footage even on a Windows PC–it just can't *output* ProRes files (which in practice is usually not an issue).

Sony created a camera system specifically designed for "vloggers" called ZV-1. The reviews are pretty good. Sony apparently sent out a lot of free samples (somehow, they missed me). However, some reviewers remarked about the short battery life, a good but non-interchangeable lens (that doesn't get wide enough for a selfie unless you are on a stick), and a sensor that gets rather warm outputting continuous video.

Sony's ZV-1 DSLR was specifically designed for content creators

True, the battery life is reported to be about an hour, but you can power the camera forever via USB. (However, that's only true if the USB port can provide the 1,500 ma required. Most PCs and laptop USB ports can't provide enough power, but a USB charger usually can.)

To prevent sensor overheating, the High setting lets you shoot at least 30 minutes continuous (the default is five, but you can change it). However, that shouldn't be a problem in most cases. Back in the day, a 35 mm film mag held 1,000 feet—only enough for about 11 minutes of continuous shooting–and we still managed to make two-hour movies. Steadicam mags only held 400 feet, IIRC—less than five minutes. The ZV-1's recording time limit is to allow the sensor to cool down, which is typical of DSLRs. None of these shortcomings are deal-breakers, IMO. And truth be told, it's based on the sensor temperature, not time per se, so factors like whether or not you are recording in-camera and ambient temps play a role here. I've recorded more than six hours straight without the sensor overheating (recording on an external device). In other words, in practice, you'll need to change cards and batts before sensor temp shuts you down.

The lens issue (only 24 mm) means that if you are shooting a selfie, the background won't be that wide. That's the only issue—selfie/background. Ulanzi makes an after-market wide-angle lens for the ZV-1 for $50 that fixes this issue. And it seems to be a function of the digital stabilization system at high res, so you'll probably get a less cropped (wider) image if you are shooting at 1080p with *digital stabilization off* than 4K with it on.

The reason you should seriously consider the ZV-1 if you are using Resolve is the plethora of recording options, including MP4, XAVC, HLG, Slog 2/3, and a host of others. It's built on the RX-100 platform, and it's similar to the Canon EOS M50 (including the price). I come from a TV background and have used Sony gear for decades, so I have a bias for Sony. Still photogs who are moving to video often have a different bias (Canon).

Still, I'm not of huge fan of using DSLR-style cameras for video. I'm old school "grab and go," but with a camcorder, you cannot approach the quality of the video of a properly tricked-out DSLR rig unless you are in the stratosphere price-wise. If you mostly shoot video for social media platforms (locked down shots/tripod), you probably can't do better for the money than a DSLR that shoots 4K. You can get a very nice complete rig based on the ZV-1 for less than a grand. Couple that with Resolve Studio, and you have a complete high-end video production platform.

Camera Settings

Before using Resolve, there are several pieces of technical information about your camera you'll need to understand and settings you'll need to configure in the camera itself. This information is probably found in your camera's manual—wherever that is. Among other things, you'll need to establish the camera's resolution, bit depth, frame rate, and compression scheme (codec). These settings are important because they not only affect the quality of the final product, they also impact how well Resolve plays and processes the footage, how long renders will take, and how powerful a PC you'll need to do all that. The Film Alliance and others have excellent videos on setting up the ZV-1.

Resolution

Resolution is probably the most important factor for achieving high-quality results—*but it's not the only factor*. It undoubtedly makes sense to shoot at 4K if your camera can do that, even if you are going to eventually output at 1080p or 720p. The files will be large, but you can do more starting with higher resolution video (like "pan and scan"). Green screen is better. Grading is better. Everything is better in high res, even if you distribute in a lower res.

An exception would be cameras that record at 10-bits in HD and only 8-bits in 4K. In that case, I might go for the HD with the higher bit depth or perhaps 4K for the higher resolution depending on how the camera does what it does. Green screen, grading, LUTs, and other effects are going to look better applied to 10-bit video rather than 8-bit, but the finer resolution of 4K might trump that. In any case, an A/B test would be worthwhile.

Frame Rate

Since almost all social media sites seem to prefer 30 frames per second (FPS—which, since the advent of color TV, is really 29.97 for technical reasons). And since you cannot change the frame rate on a project in Resolve once the frame rate has been set, 30 FPS is probably the correct choice for most social media projects.

Some folks like to shoot at 24 FPS in order to achieve "film look." But that really doesn't work because, while honest-to-god 35 mm sound film was indeed shot at 24 FPS, in the theater, it was actually projected at 48 Hz (using a two-bladed shutter that flashed each frame twice). They did that to reduce flicker (the same reason TV frames were originally interlaced).

This means that the frame rate when you watched an actual 35 mm print projected was effectively 48 FPS—not 24 FPS. But it gets complicated. Film was exposed at 24 FPS which adds a slight motion blur due to the slightly longer exposure, and theater audiences got used to that. But regardless of the camera's frame rate, you have no control over the FPS your platform uses. Furthermore, most monitors refresh at some even higher rate (60 Hz, 72 Hz, 144 Hz) anyway. Nothing you can do about that, either. If you are going for "film look," 24 FPS is one of a dozen or so things you'd need to control to achieve it. But for video on social media platforms, 30 FPS is the standard.

Chroma Subsampling

The human eye is very sensitive to changes in brightness (in video parlance, this is the Luminance or Luma component). But the eye is much less sensitive to changes in color (Chroma)—up to a point. The economics of recording files of reasonable size dictates keeping as much Luma information as possible and, if the file size needs to be cut down, tossing out some of the Chroma information. That's what happens with Chroma Subsampling which is expressed using numbers such as 4:2:2 and 4:2:0.

What those numbers ultimately mean is super technical, but in a nutshell, 4:4:4 keeps all the Chroma information the camera creates. A 4:2:2 setting discards half of the Chroma information as it is being recorded. A 4:2:0 setting keeps only one-quarter of the Chroma information. All for the sake of saving file space.

4:2:0 is what most camcorders record. 4:2:0 is what YouTube, Vimeo, and other platforms distribute (as of this writing). But for some effects and processes in Resolve (such as Chroma Key and color grading), 4:2:0 is not going to work as well as 4:2:2 (or higher) if that's available to you.

However, more pixels equals more Chroma information over the same area, so a 4K camera with 4:2:0 video will capture more Chroma information (finer detail) than a 1080 HD camera does at 4:2:2—I think. Of course, 4K 4:2:2 is even better.

The reason more Chroma info is better is because it's hard to produce a clean green screen shot after most of the Chroma information has been thrown out—not impossible, just harder. Many processes such as color grading that occur inside Resolve also benefit from having as much Chroma detail to work with as possible, so shoot with the highest Chroma Subsample rate your camera allows. Again, shoot a test with your gear and see for yourself.

In-Camera Compression—Codecs

A major setting that affects both workflows and render times, as well as quality and file size, is the codec your camera uses to compress video (before it's written to an SD card or other media). (A "codec" is software used to compress raw video and create a particiular type of file. You've undoubtedly seen these referred to as H.264, AVCHD, XAVC, ProRes, etc.)

Most consumer and prosumer camera manufacturers use very high compression codecs so they can pack more video on an SD card of a given size. That's because most consumers would rather buy a video camera that can record six hours of so-so video than 30 minutes of very high-quality video. IMO that's a terrible parameter to build a camera around because higher compression (making the file size smaller) means lower quality. It's that kind of thinking that got us VHS instead of Beta.

But the choice of codec doesn't only affect the file size and quality. *The codec choice plays a major role in how Resolve will have to handle the footage during the edit.* Many of the issues getting Resolve to play footage smoothly on a less-powerful computer are largely due to the way the camera's codec compresses video. So choosing the "right" codec (assuming your camera even offers you a choice) will impact how well Resolve works during the edit and how powerful your computer will need to be to run Resolve. (To dig into this further, Gerald Undone's YouTube channel has excellent explainers on codecs and other camera features.)

H.264/265 Codecs

Camera manufacturers choose codecs because of licensing agreements (or lack thereof) or for a variety of technical or economic reasons. For years, H.264 was the "standard" codec of choice. Sony's naming convention is different from everybody else's, and Sony calls it AVC. Sony's XAVC is an upgrade from H.264 but is built on it nonetheless.

With 4K, H.265 is becoming the new standard. If you have a choice, H.265 will reduce the file size substantially (which can be important if you are shooting 4K or higher). Qualitywise, the footage I've seen shot in H.265 looks no worse than H.264 despite the much higher compression. Like everything else, what you get often depends on little details of how each codec has been implemented and what tradeoffs have been accepted in order to get there.

There are several videos and blogs that suggest H.264/265-based codecs, such as Sony and Panasonic's AVCHD/XAVC, can't be edited in Resolve without first transcoding using some third-party software ($), but that's not the case. Resolve itself can convert your camera's footage to an easier-to-use codec such as DNxHR (or ProRes if you're on a Mac) via Optimized Media, Proxy Media, or Media Management. How and when to transcode is covered more extensively in the chapters on setup and Resolving Performance Issues. You could also use an external recorder which will give you even more codec options. And you may still have to use HandBreak or similar transcoding tools if you are importing stock footage in some weird format.

More Codecs

If your camera gives you the option, you will be better off using a codec designed for post-production, such as ProRes or DNxHD/HR, than the H.264/265 variants. External recorders typically offer more codec, bit depth, and Chroma Subsample options. Most cameras don't have a ProRes option because of the cost of licensing it from Apple. On the other hand, many standalone recorders, such as the Ninja V, do ProRes–yet another reason to use an external recorder.

The video quality of these codecs is excellent. However (you knew there would be a "however," didn't you?), shooting in ProRes or DNxHD/HR will quickly fill up an SD card or SSD and may be impractical unless you really need the image quality and performance improvement. And even so, you may still need to transcode to lower resolution proxies (for the edit) that your PC or Mac can handle easier.

Codec Settings

For the best quality your camera is capable of, use the highest quality compression setting your camera has (highest bit rate or largest file size). You'll have to read the manual to determine what those options are. Compression settings are typically measured in megabits per second (Mbps). I have a Sony camera that records at 24, 17, 9, or 5 Mbps. Obviously, there's more detail recorded at 24 Mbps than at 5 Mbps. However, more is not always better. I have another Sony camera that records at 100 Mbps as well as at 60 Mbps, but testers say there is not really any discernable difference in picture quality. Still, almost always, bigger is better.

If you have a choice between 8-bit and 10-bit video, likewise go for the bigger number. With 8-bit video, there can only be 256 values for each RGB channel. With 10-bits, that number is a whopping 1024 (only 2 bits difference but 4X the detail). There are a lot of effects and processes in Resolve that will look much better applied to 10-bit video rather than 8-bit.

External Recorder

Atomos Ninja V External Recorder with Live Scopes

Recording the HDMI output from the camera to an external SSD recorder such as the Atomos Ninja V offers many advantages. The Ninja V isn't cheap, but large, fast SD cards aren't cheap either, and you'll probably be going through them like chiclets. The Ninja V is certainly not as expensive as it looks when you consider the cost of all those high-capacity ultra-fast SD cards.

With the Ninja V, you will have more (and better) options than the camera's built-in formats allow. You'll also get live scopes (essential for proper lighting), a bright 4K monitor, and many other features. With an external recorder, it is possible to record in 10-bit ProRes 4:2:2 on a camera that only allows 8-bit H.264 4:2:0 video intrinsically (if the HDMI port allows).

Unfortunately, many camera manufacturers are mum about the actual specs of their camera's HDMI output. My guess is they don't want you to do this. The major brands seem to make essentially one prosumer camcorder and create different models with different price points depending on what recording options they make available to you.

Having your own external recorder circumvents that pricing scheme. But generally, the HDMI or SDI outputs from a camera do not have the same Chroma Subsample restrictions and none of the format restrictions the camera's internal recording system has.

It may be that getting an Atomos Ninja V is a cheaper and better upgrade than getting a new camera. And you can still record directly on the camera for backup (or to create proxies). Or you can forego recording on in-camera SD cards altogether–which improves battery life and reduces overheating–either of which can extend the camera's allowed recording time.

The Ninja V will let you record in ProRes LT, HQ, 422, and now ProRes RAW. On the DNx side of the house, you can record in DNxHD or DNxHR at various levels of compression. DNx codecs are very similar to ProRes, but even if you are on a Windows PC, I'd still advise shooting in ProRes. If not, DNxHD is fine for HD, while DNxHR is preferred for 4K. While the Ninja V doesn't come with DNxHR in the box, you can activate it (and ProRes RAW) for free. However, on a Win 10 PC, you still won't be able to export in ProRes, but at that point, it hardly matters.

If you do use an external recorder, set it up so that it records whenever the camera's timecode changes. That way, it will automatically go into record anytime the camera is recording—no second record button to push (or forget to push).

Transfering footage from an external recorder's SSD to your PC via USB can take a while. If your PC has a Type C USB port, use that for transferring (it's more than 10X faster than the standard USB ports).

Dynamic Range

All image sensors have limitations—video, film, even the human eye. Without getting too deep into the math, suffice it to say that more dynamic range is good. The more dynamic range your camera has, the more realistic the picture will be. The human eye, under ideal conditions, can perceive detail in two objects at the same time, even when one is *a million times brighter than the other*. Film and many high-end DSLRs are pretty close to that, while most prosumer cameras don't quite measure up. (I'll spare you the numbers—you can Google this if you want the nitty-gritty.)

Typical video cameras record brightness in linear variations—each step from the darkest to the lightest is more or less evenly spaced. This characteristic is called its "Gamma," and it's more or less a straight line (slanted but straight). The problem with this equanimity is that it ends up giving more steps in the mid-tones where it's not really needed than at the upper end (high brightness areas), where it would do the most good. That's because exposure is inherently nonlinear (and why f-stops are numbered in seemingly odd jumps).

To provide more detail in the highlights, engineers started fiddling with Gamma—making it curvy (and shifting some steps in the middle to the upper end) so that it's more of a curve than a straight line. The math used to generate the curve's shape results in a logarithmic (nonlinear) scale which is why this technique is commonly referred to as "log."

With many newer cameras, you have the choice of shooting with a standard (linear) profile or shooting in one of the log profiles. When you first see log footage straight from the camera, it looks flat because it has very low contrast. To fix this, the brightness values need to be remapped. Resolve does this by applying a look-up table or LUT to each pixel value. Basically, Resolve looks at the value of each pixel and refers to the LUT (which you specify) to change the recorded Luma values to more appropriate ones.

The results can be stunning—or sometimes not so stunning—it depends. Exterior scenes usually benefit from shooting log and applying a LUT because there are usually more extreme highlights and shadows in the natural world than in a properly lit studio or other interior location. For this reason, it's common to use a log picture profile when shooting outdoors and a standard picture profile when shooting indoors.

There are many folks on YouTube extolling the virtue of recording everything in log and applying a LUT (often to achieve a "film" or other esthetic look), but you may not notice the difference or even like the results if you are recording in log and only have an 8-bit video camera unless you really need the extra stop of dynamic range in the highlights. Shooting in log and applying LUTs works well with 10-bit video and less well with 8-bit. Of course, nothing is absolute. Sony's Slog2 and Slog3 profiles work surprisingly well with 8-bit video if you also use Sony's LUTs.

Which to choose depends on what you are trying to achieve and what grading and effects you intend to apply in post. Shooting 8-bit log will cost you some loss of sharpness and some added noise (particularly in the dark areas). That may not matter if you absolutely must capture detail in the extreme highlights. But generally speaking, if you want to shoot log and use LUTs, it's better to record in 10-bit 4:2:2 if available. Film naturally has a Gamma similar to log. For this reason, if you are trying to emulate film, shooting log helps achieve that.

Picture Profiles

Cameras often have a setting called "Picture Profile." Supposedly these profiles let you shoot for a "film look" or some other stylized effect. Whatever they call it, it's an in-camera modification of the video before it ever gets recorded. And, in most cases, these "enhanced" settings are not nearly as good as you can get by shooting straight video using the "standard" setting (or one of the log options) and processing it through Resolve. You can probably achieve a much more pleasing film look shooting log video and using Resolve filters and effects than any camcorder can deliver via its pre-fabbed "Cinema" setting. At least that's my experience. However, Joe at The Film Alliance on YouTube shoots a lot of really great-looking footage using various in-camera "Cinema" settings (and shows you how it's done).

If you do choose to use a Picture Profile setting, keep in mind that your camera's monitor may not be telling you the whole story. Shoot test footage, import it into Resolve, and view it on the edit monitor. Some camera viewers/monitors allow you to add a LUT on the fly which corrects the log footage being fed to the monitor so you can tell at a glance what the footage will look like once the LUT is applied.

Stabilizer

Most camcorders have a "stabilizer" setting that can help remove some of the camera shake when you are handheld. Unless you are shooting Iditarod handheld, my guess is that your viewers expect a stable horizon. Probably a good idea to have the image stabilizer on.

Optical stabilizers should probably always be on when shooting handheld. But when you also have the *digital stabilizer* on, you will lose 10% or more of the picture area (it has to be thrown away on the fly in order to keep the picture more or less "stable"—so basically, it's sort of a dynamic digital pan/tilt/zoom). Optical stabilization is generally always OK, while digital stabilization is generally not OK (IMO).

If you are shooting a lot off the tripod, it's best to use a mechanical stabilizer of some kind. There are one and two-handed stabilizers that use gimbles and weights or more expensive ones that have 3-axis gyros and motors, and all are more affordable than in the past because cameras are so much lighter. The original camera stabilizer, Steadicam, was the number one cost on some shoots back in the day, but that was when even "lightweight" cameras weighed in at 30 lbs. or more. Much easier and cheaper to achieve dynamic stability with gyros and motors when the mirrorless DSLR is less than a pound (and it probably is).

A down and dirty technique that works surprisingly well is to put the camera on a short tripod or selfie stick and hold the camera upside down. Of course, you'll have to flip the image in post, but suppressing camera shake is much easier if the weight is at the bottom rather than at the top. I got that clever idea from Joe at The Film Alliance.

Other Settings

On some cameras, there is a "Skin Tone Detail" setting that electronically suppresses the detail (softens) in areas the camera deems to be "skin." There may also be an Edge Detail setting that electronically "enhances" apparent sharpness by emphasizing the edges of objects. But Resolve has more sophisticated image enhancement plug-ins that work much better. It's generally advisable to shoot straight or log video and apply the more capable Resolve plug-ins later (if necessary) than attempting these corrections *in-camera*. That's particularly true because in-camera modifications to the video can't be undone.

Depending on the camera, it may have other settings as well. It pays to figure out what they do. Some of the camera manuals I've read are pretty cryptic, and the results promised are often marketing BS, so again, I suggest shooting test footage and seeing for yourself the results of these various settings (and watching Gerald Undone's explainer video for your camera).

Focus

It goes without saying that the main subject should be in focus—even if you want a softer effect in the final product. Softening an image is easy—Resolve has scads of options for that. But unsoftening an out-of -focus image is beyond our technology at present. Sharpening tools do exist, but generally, these tend to enhance the edges of objects to fake a sharper look. Sharpening can look cartoonish if applied with too heavy a hand.

Autofocus can help in this regard, but some autofocus systems are better than others. For many locked-down shots, (talking head) manual focus is usually better. Sometimes a simple wave of the hand will start the autofocus hunting, but the newest phase detection autofocus systems, such as Sony's Fast Hybrid AF, don't have that problem. Most autofocus systems now let you choose whether or not the subject is a human or what parts of the frame should be used for focus. And these settings can help keep the autofocus system from "hunting." If not, go to manual.

When using manual focus, it can be difficult checking focus on small camcorder monitors. To help, some cameras have a focus assist feature that zooms in or exaggerates edge detail to help you see and achieve sharp focus. Typically, the focus peaking feature doesn't show up in the recording and can be left on all the time. Check your camera's manual.

BTW, to manually focus a zoom lens, zoom in *all the way* on an object that has lots of fine detail (such as the talent's hair), focus, then zoom back out to the shot you want. Don't zoom to the correct shot and then focus. It doesn't work that way.

Depth of Field

The Depth of Field (DoF) is that area in space where objects are in focus. Outside the DoF, objects are out of focus. The DoF can be narrow or wide—depending on various camera settings and lighting.

When you are talking to somebody, your eye/brain visual system focuses on the person, and the background (if you notice it at all) goes soft/blurry. Motion picture film cameras perform a similar trick naturally (where talent is in sharp focus, and the background is soft). With TV cameras, that's often not the case. For this reason, film tends to match what your eye/brain does naturally, and we tend to think that film is "better" than TV in that regard. The reason is the size of the image sensor. With most film, the image "sensor" is 35 mm. With many TV cameras, the image sensor is only 12 to 17 mm. With a smaller image sensor, everything tends to be in focus. However, DSLR-type cameras have much larger sensors, so you can focus on the talent, and the background will be soft or out of focus, just like film.

Of course, all that depends on the f-stop. If the aperture is stopped down to, say, f-22, the DoF will be huge, and everything will be in focus no matter what. So to achieve a narrow DoF where the talent is in focus but the background isn't, you may need to adjust your lighting, shutter speed, and ISO such that your aperture is larger (f-2.8 or so) to reduce the DoF.

Lighting

Since video is a visual medium, it really all starts with lighting. There are tools in Resolve to improve the look of the footage you shot—to correct color and contrast or to fix shadows and highlights. You can even completely relight a scene you've already shot. But that takes time, and it takes talent. It's better to light properly before shooting so more time can be spent making improvements rather than fixing flaws.

There are lots of sources for information about lighting for TV and film, three-point lighting, the zone system, the exposure triangle, and all that—too much for a section of an introductory book to cover. Nevertheless, I strongly encourage you to learn about lighting for video if you want to get the best quality images from your camera.

It doesn't take a lot of money or fancy equipment to get the lighting right. I've lit sets using construction lights bounced onto Styrofoam sheets (painted with white ceiling paint). I've shot floods into draped parachutes for soft light. Lighting is what it looks like on the monitor, not what it takes to get there.

You can get a decent (but fragile) softbox kit from Amazon for around $100 or so. You can learn the basics of three-point lighting from an hour or so YouTube tutorial. If you are shooting a talking head using three-point lighting (and why not?), it's a good practice to have a ring light or some other on-camera soft light pointed directly at the talent. A little fill coming directly at the talent on-axis brightens the eyes and helps reduce dark circles.

Good lighting is easier to achieve on an interior set because you (usually) have control over what's on the set (what the brightest and darkest objects are) as well as all of the light sources. Exteriors are tougher because you have a lot less control over illumination levels and what's in frame. I usually bring reflectors (metalized or white boards) for exterior work to help reduce lighting extremes. And I use ND (neutral density) filters a lot.

Good lighting can't turn a $1,000 camera into a $25,000 camera, but it can turn a $1,000 camera into a $3,000 camera. Or, I should say, bad lighting can easily turn a $3,000 camera into a $1,000 camera.

And lighting is one of the cheapest fixes there is. You can up your lighting game for a hundred dollars, but getting the next-level camera might cost several hundred or even a thousand. If you are really serious about improving your video quality, start with the lighting.

If at all possible, set up the lighting at your shoot location and shoot some test footage. Bring it into Resolve (on a laptop) and look at it on the scopes so you don't get any nasty surprises.

Viewing test footage on a decent monitor before you shoot is also a good idea. This is particularly important if you are using location lighting (overhead fluorescents, Mercury vapor) or mixing location lighting with your lights. Fluorescent lights often have a greenish cast or other anomalies. Mercury vapor lights are the worst. This is another of those things you can fix in post with Resolve, but it's much better not to have to deal with mixed lighting issues. And if you have to deal with mixed lighting issues, it's much better to do that with lights before shooting than with software afterward.

When lighting a PowerPoint presentation, you need to decide what's important: the talent or the screen? Even with a fantastic camera, you cannot correctly light and shoot talent standing beside a typical front-projection screen. My preferred method is to light and shoot the talent (avoiding the screen entirely) and later edit in jpegs made from the ppt presentation. It's not so much an issue with the camera as the projector. There are front projectors (Barco Laser) that can be shot along with the talent, but those are not frequently encountered.

More Camera Settings

Unless you want to do a deep dive into all this, I'd recommend letting the camera handle exposure through its auto Exposure Mode system using *Shutter Priority* with the shutter set to 1/60th. That setting also reduces flicker from fluorescent lights (set it to 1/50 in the UK). Outdoors, shooting log, you may not be able to drop the ISO enough to prevent overexposure even with the ND filter on. In that case, you'll have to up the shutter speed or switch to Aperture Priority or even Program Auto. But how different cameras handle exposure is different (at least among brands), so read and understand your camera manual's advice.

Exposure

The brightest image Resolve can deal with has a numerical Luma value of 1023. Black has a value of 0 (giving a total of 1024 steps because 0 is a step). Where your video falls within that range depends on the reflectivity of the objects in frame, the amount of light available, the aperture (iris) setting on your camera along with the shutter speed and ISO (sensor sensitivity).

Shutter Speed

Setting the shutter speed for video is easy–just make it twice the frame rate (for 30 FPS, set it to 1/60th in the US or 1/50th in the UK). That's "normal." Extreme shutter speed settings are sometimes used to achieve various esthetic effects. A very slow shutter speed can help produce a "dreamy" image, for example.

ISO

ISO is basically the same thing as film speed. Fast film requires less light and vice versa. However, film speed is determined by the chemical composition of the film stock, and it is difficult to change the ISO of film. In fact, with film, you simply set the ISO on the camera to whatever the film can says it is.

However, most video cameras do allow you to adjust the ISO–*changing it*, not merely setting the camera to it. Generally speaking, you want to set the ISO to a low value because the higher the ISO, the more video noise the sensor generates. And extreme ISO settings may require you to set the shutter speed and aperture to some odd combination that's visually unappealing. Once the shutter speed and ISO have been set, the only exposure adjustment you'll need to make regularly is the aperture.

Aperture

You do not want to shoot with camera settings that record very bright areas at more than 1023 (seen on the Resolve scope as lots of junk piling up at the top of the scope's screen). The camera will respond by "blowing out" the image (engorged white blob with no detail).

I'm sure you've seen this sometimes when there's a sun-lit window behind the talent. If the camera lacks the dynamic range to capture such extreme lighting differences, the window will have no detail. It's just a fuzzy white blob. Ditto for clouds and other extremely bright areas. Outdoors, you may even have to add an ND filter to lower the brightness even more (but most cameras now have HD filters built-in). You could also lower the camera's ISO even further.

If you take the aperture down a stop or two to accommodate the bright areas (allowing you to capture detail in the highlights), the darker areas (shadows) may go full black—no shadow detail plus some noise. While that's not desirable, it is usually better than blowing out the highlights. Unless you are going for some particular aesthetic, you almost always want to capture your camera's maximum dynamic range without blowing out the white areas while keeping shadow detail in all but the darkest areas. (Shooting log helps retain detail in the highlights outdoors.)

While it's important to consider the highlights and shadows, at the end of the day, *the proper exposure of faces* is usually more important than being able to discern detail in clouds, for example. If possible, you want to properly expose your subject's face and still have detail in the brightest and darkest areas. Set design and studio lighting techniques are specifically designed to let you achieve those results. Having a camera with lots of dynamic range also helps, but sometimes, hitting all three targets simultaneously, particularly with exteriors, is all but impossible.

Whole books have been written about attaining proper exposures. The acknowledged master is photographer Ansel Adams, and you won't go wrong adapting his "zone" exposure system to video. Dianna Gladney on YouTube has excellent videos for beginners about the "exposure triangle" (and other subjects). But whatever approach you take, you should always shoot a reference shot of a 0% black object and a 90+% white object with every lighting or scene change. Then, in Resolve, you can make adjustments based on these two references (instead of just guessing) even if you don't have a fancy-schmancy calibrated monitor.

An *X-Rite Color Checker* would be more accurate, but it's $80 on Amazon. (Although there are knockoffs, many are simply printed four-color-process cards and are useless.) I don't use a color checker because the blackest color chip (2D) is not as black as a box (3D) you can make yourself (see below). And by using paint store ultra-white chips, you can replace them for free when they get dirty.

> **ASIDE**: At one time, the top of the Kodak film box itself was used to set exposure–the box was intentionally printed to be 18% gray. Photogs still use 18% gray color chips for portrait work, while TV types tend to use ultra-black and ultra-white. If you want to learn more about lighting and exposure, Parker Walbeck has excellent YouTube videos about shooting interviews, presentations, narration, and such like. Copy his techniques, and your results will be excellent too. BTW, to make a reasonably accurate 18% gray card, use the color selector in Photoshop and set saturation to 0% and brightness to 82% and print out on ultra-white paper (or whatever you've got loaded).

Reference Black and Reference White

Generally, modern cameras have fairly linear Gamma set to a standard (Rec 709). That standard assumes the user has an LCD screen or some other monitor also set to Rec 709 (probably true).

To get the most dynamic range out of your camera, a jet-black object (<1% reflectivity) should be slightly above 0 on the waveform monitor, and the brightest natural object (~90% reflective) should be about 896 (and nothing else hitting 1023).

The old-school way of setting a video camera's exposure is to buy (or make) a TV reference card (which is really a box because you'll need some Z-axis depth for the black reference). So find a smallish cardboard box (Amazon has loads of them). Cut a hole in the box about the size of a playing card. Spray the interior of the box with flat black spray paint. Obtain a playing card size color chip from the paint store of the whitest white you can find. Both Lowe's and Home

Depot have an "ultra-white" paint that is used as a base to which no colorant has been added. Get one of those color chips, preferably in a matte or flat. (Foam core also is very, very white.) Attach the white reference chip to the box with double-sided tape.

A DIY Black and White Reference

Unless you are doing something extreme, there probably won't be any natural object in frame that is brighter than the ultra-white chip—at least not indoors. There may be glints from specular surfaces, but those don't matter (no detail in glints). Outdoors, you can frequently see sky or lights that are brighter than the 90% white chip—we'll get to that situation later. And nothing will be blacker than the black box.

Scopes

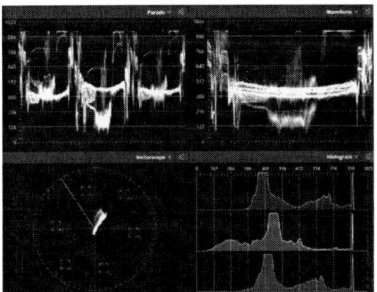

Scopes include Waveform, Parade, Vectorscope, and Histogram

There are a variety of waveform monitors (aka scopes) available in Resolve that will show you precisely the values of Luma, Chroma, even each individual RGB channel. Most cameras don't have scopes but offer zebra stripes or a histogram as a substitute. External recorders often have scopes which is another reason to use them. However, once the footage has been pulled into Resolve, you can use its scopes to adjust highlights, mid-tones, shadows, and even white balance.

Scopes are particularly useful when setting up a green screen. Typically, for green screen, you want the green screen's Luma value somewhere in the middle (around 512) and the green line fairly flat left to right (meaning the lighting on the green screen is even). The only way you are going to know if this is really happening is to light the green screen (and the foreground set), shoot some video, and look at it on the scopes.

Zebra Stripes

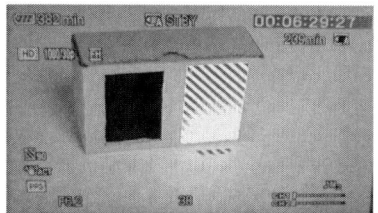

Zebra Stripes 90

A live scope is ideal for setting exposure. But most cameras don't have that feature. On the other hand, most do have zebra stripes. If your camera has zebras, use them to set the exposure. Keep in mind that no sensor or film stock has infinite dynamic range, so exposure is always a compromise between blowing out the highlights (bright details are lost) or crushing the blacks (no detail and lots of noise in dark areas) while properly exposing faces (or whatever the main object is).

If you are shooting outside, preserving details in the brightest areas is probably desirable (in some shots, the sky is half the image). If there are people in the frame, adjust the iris such that faces are properly exposed, even if that means the highlights are blown out, or the dark areas have no discernable detail. Even the human eye cannot handle extreme brightness ranges. But interior lighting usually has far less dynamic range than outdoors, so proper exposure is less of a compromise.

For outdoors, set zebras to 100+. Adjust the iris (and ND filter if necessary) such that zebra stripes are just visible in the brightest areas (sky/clouds), then back off. Under extreme conditions, faces may be underexposed, and dark shadows may lack detail. Depending on what you are trying to do, you may have to compromise—overexposing the bright areas or underexposing the dark areas so that faces are correctly exposed.

Indoors there are more options: (1) Set zebras to 90 and adjust the iris until the white reference chip shows some zebra striping, or (2) Set zebras to 70 or 80 and adjust the iris until the *talent's face* shows some zebra striping (forehead, cheeks). And, of course, you need to consider the talent's skin reflectivity—darker skin needs more light, and pale skin less. The idea is to set the exposure so that none of the detail in a person's face is lost.

There are several theories about how to go about setting exposure using zebras, and settings can vary by manufacturer, so check your camera's manual. Most of the time, I'm shooting on a well-lit set where I can control the talent lighting as well as the foreground/background objects and lighting. I set zebras to 90 and adjust exposure referencing a bright white chip. There's no chance something else will be brighter (on my set), and because of the lighting, skin tones will be properly exposed too.

Histogram

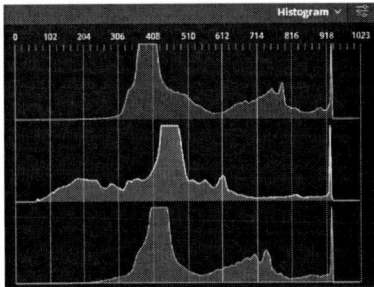

Histogram displaying a "bell curve" (use your imagination)

Many external recorders also have scopes so you can see the Luma levels in a context such that you can figure out (spatially) where the offending bright/dark areas are in the shot. You can do that with zebras, too. But some cameras only have a histogram display. Unlike scopes and zebras, a histogram does not provide information about "where" the brightest/darkest areas are.

Instead, it gives the percentage of pixels at a particular level (0%/black on the left all the way to 100%/white on the right). While the histogram is not as useful as a scope or even zebras, it can be helpful once you learn how to read it. Unless you are going for a particular esthetic (like day for night), the pattern should be more or less a (jagged) bell curve with very little pile up at the extreme left and right edges.

Lift/Gamma/Gain

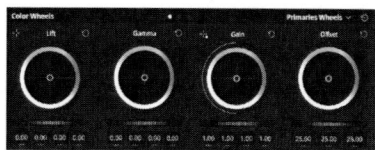

Lift, Gamma, and Gain Controls

Later on, when you are editing in Resolve, you'll have access to the Color page, where you can use scopes and adjust the Lift, Gamma, and Gain. These are the terms Resolve uses for the darkest areas (Lift), the mid-tones (Gamma), and the highlights (Gain). Generally speaking, you adjust the Lift so that the darkest black areas are just above 0 and set the Gain so that the brightest object (90% reflective) is around 896 and nothing is hitting 1023. These two settings also affect the middle Gamma adjustment (and vice versa). (Broadcast standards are a bit more restrictive and would have you set the black level at ~64 and the max white level at ~940 or so.)

"Grading" a shot means adjusting the Gain, Gamma, and Lift so that there is some detail in the brightest and darkest areas and the Gamma (mid-tones) looks natural. Grading also includes correcting color shifts, white balancing, and adjusting Chroma (hue, saturation).

Grading is often used to make shots match so they don't look like the scenes were shot on different days, under different lighting conditions, with different cameras. I've shot material at different times of the same day that, when juxtaposed in the edit, didn't even look like they were shot in the same decade (grading fixes that).

You will need an accurate visual reference for what your camera sees as 0% black and 90% white so that you can make these adjustments in Resolve. Grading is covered in more detail later in this book, but to make grading doable for non-colorists without a calibrated monitor, it helps to have shot a black and white reference (for each scene/lighting set up) that you can use in post to set accurate black and white levels.

In many situations, properly exposing the face/skin will mean the highlights or shadows will lack detail—the shadows are too dark ("crushed") or the highlights are too bright ("blown out"). One of the main reasons people are flocking to Resolve is the ability to take a so-so shot that lacks detail in the shadows or highlights, and apply color and exposure correction, and end up with a result that looks like it was shot on a much better camera in much better lighting.

White Balance

White balance (WB) the camera with a clean whiteboard or paper. We've all used typing paper or whatever was handy to color balance. There are cards made for that purpose, but I find they always get dirty despite best efforts. A good practice would be to WB the camera using a white card or paper, then shoot a few seconds of footage of the black/white reference in the actual lighting that will be used (typically as close to where the talent's face will be as possible).

You will probably notice when looking at your footage on Resolve's scopes that your camera's white balance is not perfect. (The red, green, and blue lines that make white are not perfectly aligned.) If necessary, you can use the color controls on the Color page to fine-tune white balance.

WB Eyedropper

In fact, on the Color page (bottom left) is an eyedropper you can apply to a white area in the clip, and Resolve will white balance to that. (The "A" next to it is an *automatic* white balancer.) Notice that Resolve's WB is better than your camera's WB. But I mention this simply to let you know where these controls are and how they work. No need to put too fine a point on it because when you start grading, you'll be overriding the WB anyway. WB is really there just to correct gross color shifts.

Setting up the ZV-1 for Resolve

After re-reading my own review, I decided to get a ZV-1 (from B&H). The camera comes with a quick start guide that basically shows you where the buttons are, where the battery goes, and not much else. Detailed information is in a 555-page web guide (there's also a PDF version you can download).

Five hundred-plus pages printed out means the manual weighs more than the camera itself and is about 4X as big. It's good, I suppose, that there are so many customizable features that the ZV-1 needs a manual this gargantuan. But such tomes can make setup boring and time-consuming. So my approach was to write down a list of the few critical settings I wanted to review/change and check those off as I went through the setup procedure (resolution, frame rate, codec/file size, ISO, shutter, log picture profile). (There are over a hundred user settings, and these settings need to be understood and set properly. Film Alliance on YouTube has several videos about how to set up the ZV-1 depending on what you are trying to do with it.)

For now, I want to shoot 4K using Sony's Slog 2 Gamma profile. I know the file will be XAVC. I'm not a big fan of that format, but that's the only practical option, given the other settings. The ZV-1 records this resolution at either 100 Mbps or 60 Mbps. While bigger is usually better, I've read in lots of reviews that no one seems to be able to discern any difference in quality.

Setup required looking through the manual in PDF form (for its search capability) and wasn't too bad. (DB Vares has a series of excellent ZV-1 videos on YouTube.) The camera's menu/controls are more than adequate if you are not in a huge hurry. I was able to skip a lot of the manual because the front half is about using the ZV-1 for still photography.

I didn't count how many control settings the ZV-1 allows, but there must be more than 100. One handy feature is the camera lets you create a custom menu (actually several of them) so that when you hit the Menu button, there are six functions (per menu page) you can control. I have White Balance, Picture Profile, Record Setting, Zebra Setting, Format (erase), and ND filter set up on a custom menu, so I don't have to wade through Sony's extensive lists to find these controls.

The built-in camera mic (three capsules) picks up too much noise and room coloration for my taste. (A tiny "dead cat" windscreen is included that works surprisingly well.) So in an emergency, you could grab just the camera and go. The ZV-1 will (apparently) not let me record two mono tracks (one for safety at a reduced level). Instead, when I need to do that, I'll use an external audio recorder. Another option would be to use an obsolete camcorder just to record the audio. You don't have to sync these safety tracks because you are just pulling a word or two here and there and replacing like with like.

Importing media from the ZV-1 into Resolve is a breeze. You just plug the camera into your PC (assuming you installed Sony's app), and a folder with your clips appears.

The picture quality, given the 1" Exmor sensor, is stunning compared to a Sony prosumer camcorder that cost more than $4K only a few years ago.

Talent

It's a good idea to do some camera tests of makeup. What looks great in a mirror may not achieve the desired results on camera. In general, at least use a translucent powder to tone down any shiny spots. And keep a lint roller handy to pick up lint and bits of hair or dandruff—particularly if the talent is wearing solid or dark colors. You can remove these offending bits with the patch tool, but a lint roller is faster. For Chroma Key, holding the hair in place with a gel or spray is preferable to the Einstein look. I always keep a lint roller, neutral translucent powder, a can of hair spray, a hairbrush, antiperspirant (see below), and a disposable razor in my camera bag.

Shooting for Editing

If possible, shoot everything twice. At a minimum, that will give you a safety. If the master shot is medium, shoot the safety as a close-up and vice versa. When editing, you can then switch back and forth between shots without having to deal with jump cuts. Jump cuts can be covered with B-roll if available. If not, Resolve also has a Smooth Cut plug-in (in Video Transitions) that sort of melds two clips together to reduce the impact of the jump cut. But it's much better to be able to simply cut back and forth between two visually dissimilar shots.

The standard setup for interviews has long been to shoot a master (usually wide) and then shoot the interview subject "over-the-shoulder." Even with one camera, you can fake that by starting with a wide master shot for the first question and answer. Then move the camera and shoot the rest over the shoulder. Get at least some footage of the questioner "listening" so you'll have something to cut to. Many times the questioner is "off mic" during the interview. If that's the case, mic the questioner and record them re-asking all the questions on camera.

Luma and Chroma Corrections in Resolve

After your lighting is set, white balance the camera and shoot a few seconds of the white and black reference (placed about where your talent's head will be). Then shoot your interview (or whatever). Put that clip in your Media folder.

Open Resolve, start a New Project (named "test footage" or similar). Go to the Media page and select File > Import File > *Import Media*. Select the test clip and drag it into the Master Bin and from there into the Timeline. The first time you do this, Resolve creates a Timeline which is a clip-like entity.

Click on the Color page. If you shot the footage in log, select the clip's thumbnail and right-click. Here you have options for applying a LUT to the clip. Which one you pick depends on your camera. Always apply a LUT (if you are shooting log or similar) before you make any further corrections.

Go to Workspace > Video Scopes and select *On*. You'll get four scopes. At the top right of the scopes, you can select how many. Select two. You want the Parade and Waveform scopes active. You can change the size of the scopes and move them around if they are in the way.

The Parade shows the values of RGB separately. The Waveform monitor shows you the combined values of all three (Luma). You should be able to figure out (spatially) where the white and black reference lines are on the Waveform monitor. The line nearest the top is white, while the line nearest the bottom is black.

The first thing I notice when I do this is that the white balance system on my camera is not exact. Either the red, green, or blue is dominant in the white reference. I'll either use the eyedropper white balancer or adjust the individual RGB controls.

The default controls are the Primaries Wheels. There are four Luma controls—Lift (shadows/black), Gamma (mid-tones), Gain (highlights), and Offset (overall).

Adjust the white reference line by dialing the horizontal Gain wheel up or down (actually, left or right) until the brightest object (white chip) is around 896 (assuming the white reference is, in fact, ~90% reflective), and nothing is hitting 1023 (clouds, snow, or other extremely bright areas). If your video exceeds 1023, Resolve will limit it for you in ways you might not appreciate. In that case, adjusting all three Luma controls may be in order rather than just cutting off the top. By looking at the waveform, you can also get an idea of how good your lighting is and what needs to be done about it.

You should also be able to figure out (spatially) where the black reference line is on the Waveform monitor. Dial the Lift up or down until the black reference line is just above 0. Black will likely be color balanced better than the white reference, but you could adjust RGB values in Lift if you needed to. Usually, black (real, honest-to-god black) has so little Luma that any Chroma shifts won't be noticeable. That said, sometimes dark shadows can show a pronounced color shift, but you can fix that using the color controls in Lift.

Reset Circle

You may need to go back and forth a few times adjusting Gain and Lift because they interact to a degree. FYI, the circular arrow at the top right of just about every control in Resolve resets it in case you overshoot or just want to get back to normal. And you may also need to adjust Gamma until the face/skin is right (which may require tweaking Gain and Lift again).

At this point, you've white-balanced the clip and set the basic Luma levels. The result you want to end up with is a thin white reference line (equal values of RGB) at about 896 or so with no highlights pushing into 1023 (depending on how white your white reference is) and with ultra-black slightly above 0 and nothing pushing lower than that. However, while hitting the numbers is great, at the end of the day, what it looks like on the monitor is all that counts. So respect the numbers, but don't be a slave to them.

In the middle/bottom of the screen, you'll note that you are in Curves. Instead of turning dials, you could have moved the white diagonal line, grabbing it at the right for Gain and the left for Lift. You can also adjust Y (Luma) and/or RGB values together or individually. Basically, the same function, different controls.

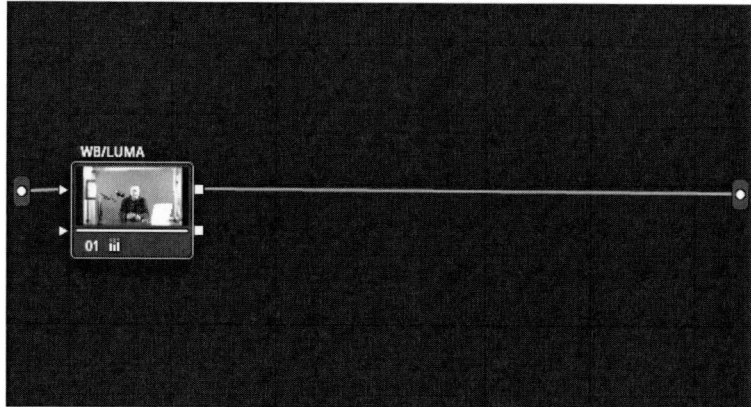

Color Page Node

I would be remiss if I didn't point out the little box with a thumbnail of your clip (outlined in red) located in the upper right-hand corner of the screen. That's one of the fabled Resolve "Nodes" you've heard so much about. You just created one and didn't even know it. Wasn't that hard either, was it?

Notice that the bottom of the Node has icons representing what attributes you added. You should have only one icon at this point (Bars). It's not as informative as a label, so right-click on the Node, select Node Label, and give it a name like "WB/LUMA." That way, when you add more Nodes, you'll know what this one is for. You can turn an individual Node off and on by clicking on the Node's label.

At this point, your camera is white balanced, color corrected, and is delivering all the dynamic range it's capable of. When you first shoot a scene, set exposure, white balance in-camera, and shoot the reference box. Do that again for any lighting or exposure change. When you import media, create a WB/LUMA node using the reference footage and copy the WB/LUMA node to any other scenes shot in the same light with the same exposure setting. How to do that is covered in the exercises.

> **MOVIOLA**: Despite what I said earlier about Moviola—the old-timey editing device—check out their current website. It's designed for budding filmmakers and has a wealth of rock-solid 4-1-1 on writing, directing, lighting, and visual effects. Moviola also has a free "Coffee Break Film School," which is cheaper than actually going to USC. Among the lessons are how to be a grip and how to operate a boom mic. Of particular note is the lesson on *Cinematic Documentary Storytelling*. The whole point of schlepping all that gear around is to tell a story using pictures and sound. This one-hour lesson will change how you think about visual storytelling.

Teleprompter

If your talent is using a Teleprompter, try to get the camera rig about 10 feet from the talent. The distance helps minimize eye shift. If your talent has not had lots of experience reading a prompter, have them practice, practice, practice. I've worked with broadcast newsreaders who spend hours each day before a prompter, and of course, they're great at it. But corporate executives and others who haven't had much experience reading from a teleprompter need to practice. Usually, the first thing I do on set (after coffee) is set up the prompter so they can start practicing while I'm assembling the lights, running power cords, setting up the audio, and all that. It's a good use of their time and helps keep them occupied. Also, you can sneak a look at the audio monitor for a more-realistic audio check.

One of the best teleprompter apps I've seen is PromptSmart Pro (IMO). It runs on the iPad or Android-based tablets. Using voice recognition, it scrolls the text as the talent reads it—no need for a teleprompter operator. And it's only about $20 (one time).

Antiperspirant

A few years ago, I was backstage in the green room chatting up the talent who was about to go on. There were about 10,000 people at the venue and five jumbotrons. About 30 minutes to air, the talent decided to dry shave some of his 5 o'clock stubble and made a deep gash across his Adam's apple. Honestly, it looked like a suicide attempt. The makeup person was called back, and she tried everything in her bag of tricks. Production assistants were scurrying around, trying anything they could think of, from Band-Aids to gaffer tape. Nothing worked. Blood continued to ooze from his throat, and the clock kept ticking. Did I mention he was wearing a white turtleneck?

Then I remembered an old barber's trick: the styptic pencil. Back before blood-borne diseases were understood, lots of men got their shave from a barber using a straight razor. Of course, there were the inevitable cuts, so the barber took out the styptic pencil (which looks like a piece of chalk) and touched it to the cut, which instantly stopped the bleeding.

It was late at night, and no drug stores were open, and anyway, I didn't think they even made styptic pencils anymore. But from chemistry classes, I remembered that the main ingredient was alum. Alum is responsible for the pucker factor in pickles. It also causes blood vessels to "pucker," tightening them up and stopping the flow of blood.

I thought to myself, "Where can I get my hands on some alum in the next 5 minutes?" Then I remembered that alum (or a more modern equivalent) was the active ingredient in antiperspirants. Basically, antiperspirant puckers the sweat glands, squeezing them shut. I remembered all this because the inventor of antiperspirant was a Dr. Jules Montenier. His company's product, *Stopette*, was the sole sponsor of one of the first game shows back in the Golden Age of live TV (*What's My Line*), and I'm am enthusiast of TV's "Golden Age.".

So, I asked the makeup person for an unopened stick of antiperspirant (I didn't want to give the talent an infection). We daubed it on the gash, she hit it with some concealer, and at 3 minutes to air, the talent headed for the stage. We watched the monitors for any sign of bleed through, but the antiperspirant did the trick.

I don't know what this story has to do with TV production except that there are all sorts of gremlins and Murphys constantly conspiring to mess up your shoot, and sometimes you have to think outside the box to get the show back on track. Also, if I didn't tell this story here, it'd never get told. So there.

Recording Audio for Resolve

Recording Audio for Resolve

Microphones

I've used some really expensive camera-mounted (shotgun) mics over the years. Sometimes that was the only way to get the audio. But a camera-mounted mic—even a shotgun— should not be your only mic. (And the mics built into the camera are worse.) I've rarely heard a camera-mounted mic at any price that's better than even a cheap wired or wireless lapel mic.

You might have seen feature films or TV series using a boom. Usually, these are covered with a "dead cat" windscreen or a blimp and operated by a technician who's constantly monitoring the sound with headphones and continually repointing the mic and adjusting levels. If you have such a person on your staff, great.

What you don't see are the numerous wireless mikes hidden in clothing or hair or behind foreground objects which are frequently used to cover what the boom missed. Even then, troublesome dialog (particularly from a boom) often gets replaced in post using Automated Dialog Replacement (ADR). In that technique, the talent watches a looped replay of the scene and re-records the dialog in a quiet studio that has absolutely no acoustic coloration of its own ("dead flat"). Later, the sound editor EQs the audio (and adds echo and other tricks) to get it to match the rest of the location audio.

Without lots of practice, a boom is hard to keep out of frame, which adds another complication. (Not a problem if the talent and boom are fixed.) If you are on location, keep in mind that booms can pick up acoustical noise from air conditioning vents and electrical noise from fluorescent light ballasts.

If you use a boom indoors, be cognizant of what it's actually pointing at. With seated talent, that's usually a table (very reflective). If possible, angle the boom mic so that it points more at the talent's mouth/chest and less at walls, floors, or other flat surfaces.

Placement

Microphones should be placed near or aimed at the throat/sternum rather than simply pointed at the mouth. Lapel mics are generally preferred. It's usually better to place a lapel mic on the shirt front rather than the collar because of head/neck movement. Always be careful not to let the talent's clothing rustle the lapel mic. Admittedly, I sometimes use a shotgun for safety, but the main mic is always a lapel mic unless that's wholly out of the question.

The further away from the talent's mouth/throat/sternum the mic is, the more room coloration you'll pick up (and the more treatment the audio will need). Obviously, too close is too close, but try to be no more than a foot or so away from the talent's mouth. I recently saw a YouTube video about using audio noise reduction. The poster's mic could not be seen in frame, so it had to be a camera-mounted mic at least six feet away from his mouth in a room full of fans running full-tilt and a mini-split AC unit overhead—hence the need for audio noise reduction...

The next best option is a desk mic. But those could have problems if the talent moves off-axis too much. Some cheaper desk mics can pick up a lot of vibration from the table or desk. Usually, I'll cover the desk with a carpet remnant if it's not on camera. Then again, I sometimes use a desk mic for its esthetic value (think Larry King or David Letterman's RCA 77DX desk mic—spoiler: those were fakes). With podium mics, the talent sometimes fiddles with the mic (nervous energy) or turns or even steps away, which causes problems.

A lot of folks use pop screens on mics to reduce the chance of popping "P"s (plosives). Pop screens do help—a little. But having the talent a tad off-axis is probably more effective. Here in Nashville, where lots of recording is done, I see audio engineers use pop screens all the time—not so much to reduce pops (they are not really that effective) but to keep the talent's spittle off the diaphragms of their expensive Neuman condenser mics. That's really what "pop" filters are for.

One additional point I would like to make concerns recording voice-overs (VO). If the talent does some of the presentation on camera and then records a VO to be used under B-roll, they should record *all of the audio* at the same location, at the same time, and using the same mic set up as the on-camera video. Otherwise, you'll end up with a sort of audio "ransom note." The SOT and the VO won't match aurally, and the mismatch will be very noticeable when edited checkerboard fashion. That can be fixed—sort of—but that's just another one of those things where it's best to get it right from the get-go.

Audio Recording

As with video that's too bright, you don't want to record sounds that are so loud that the audio system "clips" on the peaks (which happens at 0 dB and above). That results in audio problems that are hard (or impossible) to fix in post. Yet you want to record talking heads, presentations, and narration at fairly high levels. This leads to a sort of "Price is Right" dilemma where you want as much as you can get without ever going over.

One way to do that is to record the talent audio as two mono tracks. It's a good practice to put the main mic (wireless, desk, or boom) on one track and use the second track for safety—using a separate mic (on-camera shotgun). Another way to obtain a safety track with just one mic is to simply duplicate Track 1 on Track 2 at a reduced level (down 9 dB or so). That way, if the talent gets too excited, and the audio clips, you can edit in a word here and there from the safety track. Tascam has an inexpensive recorder that has a "dual" mode that lets you duplicate a mono source on a second track at reduced volume—with one button.

If you use a separate audio recorder, you'll need to be able to sync the video and audio. That's what the "clapper" is for. The third assistant AD would say the scene and take number and clap the sticks. Later, the audio playback head would be set right on the clap, and the film advanced frame by frame until the clapper just closed. At that point, the tracks would be locked, and everything would be in sync.

Another way is record yourself stating the scene and take number and then just clap your hands on camera. It's fairly easy to slide the video until it syncs with the audio. (You may need to turn the Snap tool (magnet icon) off first, so you have complete freedom in moving the tracks.) Once synced, Link the tracks, and they will stay in sync.

Most camcorders have fairly good automatic level controls. Whether you control the levels manually or automatically, aim for -12 dB peaks on the main channel and somewhat less for the duplicated safety track.

Most low-frequency rumble is associated with the powerline frequency of 60 Hz and its harmonics. Turn the low cut/high pass filter on (if your mic has one) to reduce rumble—better to not have it than have to take it out later. That option may also be available in your camera or on your external recorder. These filters have gentle slopes (6 dB/octave) and do not actually remove everything below 80 Hz or so. Resolve has audio filters you can apply in the edit that are both adjustable and more aggressive, but if you move the low/rumble filter much beyond 100-200 Hz you'll cut into the vocal quality (particularly with deep voices).

So do what you can to eliminate low-frequency noise *at the source* when you are recording audio and use a rumble/high pass filter for good measure. Before recording, *listen to the room* with a good pair of headphones with the volume turned up—way up. If there are audible noise sources, locate and turn off during recording. In particular, shut off HVAC systems (at the thermostat), blowers, fans, and motors, old-school fluorescent lights—anything that makes noise. That means mobile phones, too. Have the talent move around to catch any bad audio connectors or rustling clothing.

In small rooms or near highly reflective surfaces, use carpet remnants, moving blankets, or other dense fabrics to help reduce sound reflections from floors, walls, and nearby surfaces (desks). Large sponge-rubber blocks are the traditional sound absorbers in recording studios, but they are unwieldy and expensive. Foam packaging materials and foam mattress toppers are much cheaper and work fairly well on walls. But be advised these materials are highly flammable (Google *Providence Nightclub Fire* for more info). Professional sound-absorbing materials like Sonex are fire resistant but expensive (although the stuff looks great on camera).

Continually monitor the audio with headphones and watch levels during takes (adjusting if necessary). And listen not only to the talent but also for any extraneous noise sources like sirens, cars, trains, or planes. Keep in mind that talent typically gets louder as they get more comfortable with the process. So setting levels with an initial "testing 1, 2, 3" really doesn't work—have the talent read a paragraph or so. Either monitor and adjust the audio continuously or record a safety track at reduced volume you can cut to in places where the audio clips.

If you read a lot of YouTube comments, you'll probably never find one complaining about grading or crushed blacks or blown-out highlights. But you'll find tons of comments about audio—general room noise, coughs, mic bumps, reflections, echos, off-mic talent, clipping, audio too low, audio too high.

The number one audio complaint is music: too much, too often, too loud. It's OK to have a musical stinger in the intro, but a continuous music bed should be used rarely, if at all, and should fit the genre. Many viewers will dump out of a video just because there's continuous music under the narration. If you absolutely must have a music bed under the narration track for some esthetic reason, duck the music 6 dB or more when the talent is speaking, and EQ the music track to reduce the level in the 1-4 kHz range by another 6 dB so that the music competes less with the VO.

As YouTube videos have gotten longer and longer (to accommodate mid-roll ad breaks), audio quality has become more important. Bad mic placement, extraneous noise, mic bumps, pops, and music beds that were acceptable in short videos are much more annoying in long-form and can cause viewers to bail. When you are done with the audio, listen to it the way your viewer will (speaker system, laptop, headphones). If you mix and sweeten your audio tracks only with headphones, and your viewers will be listening mostly on speakers, they won't be hearing it the way you heard it in the mix. (I highly recommend watching a few videos from Filmmaker IQ about recording audio for video. Very good, practical advice.)

Zoom

Because of the pandemic, many presentations have been recorded where the main subject was being interviewed over Zoom or similar platforms. The interviewer recorded their side of the convo with a DSLR and a decent mic. But the interview subject was simply shot on a smartphone *using the phone's mic* and often in a room with no acoustical treatment. The results are less than stellar. And it sounds doubly bad because of the back and forth between the interviewer (good audio) and the interviewee (bad audio).

In these cases, I've found it helps to send the interview subject a digital recorder and mic. I use Zoom to help them set it up. If they are savvy, they can pull the file off the SD card and e-mail it to me. If not, I pull the SD card when they send the recorder back.

This doesn't help with the video, but usually, that's not where the negative comments come from. Most viewers will readily accept smartphone video when they won't accept smartphone audio (at least not for long presentations or when intercut with a much better-sounding audio track).

Room Tone

Room Tone is just the ambient sound of the room itself (without the talent speaking). When you are editing, if you need to remove an extraneous burst, like a cough, mic bump, or a whole word, simply cutting it out often makes the fix more obvious and more objectionable than the noise itself.

A better practice is to edit in an equal-length snippet with just the sound of the room—the Room Tone. Done that way, the fix is inaudible. I can't think of any edit I've ever done where I didn't have to add a snippet or two of Room Tone. Often, that's a better way of fixing shot noise than trying to minimize it by ducking or other techniques. It's certainly faster.

To have Room Tone available for the edit, you first have to record it. Before the talent gets started, record 10 seconds or so of Room Tone at each setup/location. Having Room Tone available also helps when editing video that has no audio with it or whenever you want to keep the video and kill the audio.

Audio noise reduction plug-ins usually need to hear what just the noise sounds like in order to remove it. Usually, it looks for noise in the spaces between words. But you can remove more noise and create fewer artifacts if you can present a clean noise source to the noise reduction plug-in using a short piece of Room Tone. This can be done by having the noise reduction plug-in "learn" the noise signature based on Room Tone at the beginning of the clip. Once noise reduction has identified the noise, you can remove the Room Tone segment.

> **CAVEAT:** Although Resolve includes Fairlight, a complete Digital Audio Workstation, I think Fairlight is overkill for most social media video projects. It's there if you need it, but I'm sure you'll find the Open FX audio tools on the Edit page more than adequate. If you do intend to use Fairlight, my advice would be to not use audio effects on the Edit page (or other pages) and Fairlight in the same project.

DaVinci Resolve Setup

Resolve lets you customize how it works by setting global preferences, workstation preferences, and user preferences, as well as preferences for individual projects. Resolve may work quite well right out of the box without you having to change many (or even any) of the default settings. However, because video editing challenges your hardware like few other tasks can, it's usually desirable and sometimes necessary to tweak Resolve's preferences to get the most from your camera and your PC or Mac.

Because Resolve is based on several stand-alone programs now under one roof, which settings control what functions (and where those controls are located) is not always straightforward. Some controls that affect simple functions are spread over several locations, while some sophisticated features have only a simple on/off switch—*go figure*.

The wonderful thing about Resolve is all of the settings it allows you to change to your personal preference, needs, or taste. The problem with Resolve is all of the settings it allows you to change. As you will see, there are some two dozen settings in Resolve that *just affect playback performance and rendering*. And most come with tradeoffs that may or may not be desirable, acceptable, or even obvious. Oh, and did I mention they often interact with each other in obscure ways?

Setting Up DaVinci Resolve

With Resolve 18, you can probably go to the end of this chapter and just use the cheat sheet provided. *In fact, most users can probably just accept the defaults in Resolve 18.* However, you will, at some point, need to make changes to the system and user preferences. So it wouldn't hurt to read through these settings, even if you are OK with the defaults. Note that there are fine points and one-off exceptions that I have avoided covering in an introduction, so if the noob advice doesn't work for you, you'll need to consult the bible (*The DaVinci Resolve Reference Manual*—it's a free download from Blackmagic Design). Also, check out the official DaVinci Resolve blog. And, of course, Reddit has lots of posts that cover Resolve's initial setup and settings.

The following settings are suggested to help achieve smooth playback and shorter renders with the fewest tradeoffs (at least you'll know what they are). Some settings depend on your camera's resolution, frame rate, and codec and what tradeoffs (if any) you are willing to make due to your PC's (GPU/CPU) limitations.

Preferences

The first thing you'll want to do when you first run Resolve is set Preferences. The following recommendations are based on using DaVinci Resolve Studio 16, 17, or 18 on a Windows 10 PC. The Mac version is very similar, however.

To get started, install and run Resolve. On the DaVinci Resolve pull-down menu, select *Preferences*. There are two types of master preferences: *System* and *User*. For the first step, start with *System*. Keep in mind that in most cases, you'll have to restart the program for the changes to take effect.

Memory and GPU

Here you can limit Resolve's access to system memory. But for best performance, it's a good idea to let it have all it wants and not run any other programs concurrently or in background. In other words, don't check the news or your e-mail while Resolve is rendering. That will just slow things down.

GPU Configuration > *GPU Processing Mode* configures Resolve to use your graphics card for renders and similar processes. Set it to CUDA if you are using an Nvidia card, or to OpenCL for AMD, or to Auto if you don't know. Make sure the drivers for these cards are the latest versions. Many of the performance issues I had with a newly-released graphics card were solved by updating the driver.

If you are using an Nvidia card, they have GeForce drivers specifically tuned for Resolve. The standard installation gets you drivers optimized for gaming. Most users say there isn't that much of an improvement, but every little bit helps.

On version 16, if you had more than one graphics card, you could check *Use display GPU for compute.* That sped up rendering because you are using both cards for render computations. (This was only an option in Studio.) Of course, if one graphics card was less powerful than the other, you'd use the slower one for display only and uncheck the box because GPU render performance is only as fast as the slowest GPU.

With versions 17 and 18, you can either *Auto* select *GPU Configuration* and let Resolve sort it out or uncheck the *Auto* box and manually control what card does what. For example, *uncheck the display GPU* if both cards are not identical or if you don't want to use one of your GPUs for render computations for some reason. In other words, if you have two video cards and they are different, don't let the slower one help with render computations. Renders will then only go as fast as the slowest GPU. In that case, just use the slower GPU to drive the display.

Media Storage

The most important thing to know about Media Storage is that the *first storage location* on the list is the scratch drive where Resolve stores all the Render Caches it creates (which is a bunch). Many render files get created during an edit; they are large, and Resolve accesses them frequently. Make sure you have your fastest/largest drive at this location. Later on, if you want, a specific project could be set to use a different drive as the scratch/cache drive (Project Settings > Master Settings > Working Folders).

Add any other storage locations you plan to access from inside Resolve itself (it makes it easier for Resolve to locate files via the UI rather than your OS). At minimum, that would be a media import folder (input), a distribution folder (output), and an archive folder (long-term storage). Depending on your OS, you may have to create these folders before you can select them (oops). Also, check *Automatically display attached local and network storage locations.*

Decode Options

If you have a graphics card that can handle H.264/H.265 hardware decoding, click the *Decode H.264/H.265 using hardware acceleration* box. It will speed things up to offload these tasks to silicon (Studio version only). All of the latest AMD graphics cards use a fast hardware decoder for these codecs, as do most Nvidia cards. Check the AMD box if you are using one of their cards.

> **NOTE**: When outputting your finished project via the Deliver page, you should change the *Video Codec Type* from Native to Nvidia or AMD to use your card's hardware encoding feature if available. The GPU card-specific settings in Preferences apparently only affect edit functions.

Video and Audio I/O

For now, just make sure your Speaker Setup matches your hardware.

Internet Accounts

If you want, you can set up the info needed to login to your YouTube, Vimeo, Twitter, or Frame.io accounts from Resolve and have it automatically upload the final output. Or you can wait till later.

User Preferences

Most of the following *User* settings are now the default in Resolve 18. If you are using an earlier version, follow these recommendations:

Under *UI Settings*, a handy box to check is *Reload last working project when logging in*, so you don't get the "Create a New Project" screen every time you start up. Now you'll be taken to your latest project every time you start.

Under *Project Load and Save*, enable *Live Save* and *Project Backups*.

Under *Playback Settings*, click *Hide UI Overlays*—which keeps your graphics card from having to constantly update onscreen panels during playback. This takes some of the load off the graphics card and lets it concentrate on rendering. When Resolve is not playing the Timeline, these overlays are reactivated and updated.

Another box to check is *Minimize interface updates during playback*. This reduces how often the PC looks at the mouse and other tactile inputs during playback.

Performance Mode runs a few simple benchmarks on your hardware and adjusts certain settings to help improve playback performance. It defaults to *Automatic*. You can also disable it or disable individual settings. Performance Mode is only active during playback and does not affect the quality of the final output. It seems that the results of the Performance Mode benchmarks are used in Resolve's Smart rendering algorithm when trying to figure out if your computer can play the Timeline while rendering on the fly or not. So turning Performance Mode off may affect Smart render behavior.

Scan the rest of the System and User Preferences in case any others apply to your situation (probably not at this stage).

When you are done, click *Save*—with the knowledge that you can always get back to the original factory default settings. *Reset System Preferences* is behind the three dots at the top right of the Preferences control panel.

Restart Resolve to make sure all the changes you made take effect.

For those who don't want to learn new things, *Keyboard Customization* lets you change the keyboard bindings to match Avid, Final Cut, or Premier. The control for *Keyboard Customization* can be found under the DaVinci Resolve pull-down menu. You can also set up Fairlight to mimic Pro Tools.

Project Settings

Gear

Now that the master System and User preferences are set, you can also set various preferences for a specific project. *Project Settings* can be accessed from the gear symbol at the bottom of any Resolve page, or go to File > *Project Settings*. The Timeline's frame rate must be set *before the Timeline has been populated with media* because it's impossible to change it later. AFAIK, that's the only setting you can't change midstream.

Presets

Initially, there are three default Project Settings presets:

(1) the settings for the *Current Project*,

(2) the *System Config* settings (factory default), and

(3) *guest default config* settings (if any).

The *Current Project* settings are the settings for the current (open) project, of course. Initially, these settings are taken from the default *System Config* preset. The Current Project settings can be edited and saved (but only apply to the open project).

The *System Config* preset is the factory default (and you can't change those settings). However, you can modify and save *guest default config* settings or create additional presets. It's a good idea to create presets based on your camera's resolution, frame rate, codec, and any other parameters that matter to you and save them so they can be reused. Do that, and you can pick the saved custom preset from the list whenever you create a new project. You could also tell Resolve to use your custom preset as the default whenever a new project is created.

To create a custom preset, make any changes you want on the *Current Project*, then come back to the Presets panel and click *Save*. Resolve will ask for a name. Here it will be useful to give your custom preset a proper name such as "1080p 29.97 fps" or whatever. The custom preset is now on the list of presets. When you start a New Project, go to Project Settings > Presets, select the custom preset you want to use, and click *Load*.

The other option is to right-click on a custom preset and select *Save As User Default Config*. If you do that, every New Project will start with the custom preset you just created.

Master Settings

Under Master Settings > *Timeline Format*, you set the Timeline resolution and Timeline frame rate. Changing the *Timeline resolution* to a lower resolution than the source footage is one approach if you are experiencing Timeline playback/render issues. You could shoot in 4K and edit in HD or less, then change the resolution back to 4K (a re-render will be necessary), and output from the 4K media.

Blackmagic claims that Resolve is "resolution independent," meaning that the Timeline resolution can be changed at any time (certain Fusion clips excepted). That's true, as far as it goes, but changing the Timeline resolution means that all clips in the Timeline will have to be re-rendered at the new resolution when you change it back. However, lots of people use this workflow to improve playback performance while they are at the keyboard and let Resolve re-render everything when they are going to be AFK a while.

> **CAVEAT**: If you set the *Timeline resolution* to less than the source footage, before outputting the project, *change the Timeline resolution* back to the original resolution. *And do this in the Timeline itself via the Master Settings*. Changing the resolution on the Deliver page is not the same thing. Changing the resolution on the Deliver page could cause Resolve to upscale your project from the reduced-resolution Timeline. I'm guessing you don't want that.

The *Timeline frame rate* should be set to the same as your source footage or to the frame rate you intend to use for the final output. Ditto for the *Playback frame rate*.

When you import the first clip, if the frame rate of the source footage doesn't match the setting in the project's presets, Resolve will ask you whether or not you want to change the frame rate to match the footage. Usually, this is a good idea. Sometimes, not.

Resolve is a stickler for accuracy, so if your footage is 29.97 FPS and your Timeline is 30 FPS, you'll get a question about changing the frame rate. If you allow the change, the Timeline frame rate will be set to the frame rate of the *first file imported* instead of using the project's presets. If you don't allow the change, the project's preset will be used instead.

Either way, the Timeline's frame rate will be established. At that point, *the Timeline's frame rate cannot be changed* unless you delete all the media files (and the Timeline itself) and start over.

Video Monitoring

This should be set to the actual resolution of your monitor (or less) and the same as (or less than) the Timeline resolution. For *Video bit depth*, set to 8-bit unless your monitor supports 10-bit, and you really need to see the video in 10-bit during the edit. Higher bit resolution will be somewhat harder to play back smoothly and takes longer to render *because two more bits per color per pixel will have to be calculated. Set Monitor scaling* to Bilinear (normally the default). Basic has been reported to be a tad faster, but I can't prove that.

Optimized Media, Proxy Media, and Render Cache

OK, here's where it gets interesting. With Resolve 17, BMD added "Proxy Media" to the options mix. So there are now at least four ways to operate at a reduced resolution (and, in some cases, using a different codec) which can substantially improve playback performance on your computer. And Resolve 18 vastly improved the handling of proxy files.

You will probably want to use some combination of these tools in your workflow. But understand that these optimization/proxy/render settings are simply designed to ensure that the Timeline plays on your computer mostly smoothly, most of the time, because when that doesn't happen, Resolve will need to halt and render. That's all these next settings are for. But they are critically important to both your workflow and your psyche. The Resolve 18 default settings here will probably work fine for most people.

Proxy Media

Proxy Media lets you work with low res proxy files using a codec designed for post-production instead of your footage's original resolution and codec. If the resolution or codec of your camera causes playback to struggle, you can transcode via Proxy Media to something Resolve (and your computer) will find easier to handle.

Proxy Media can transcode the camera's original files to a codec designed for post-production and at a lower resolution if you so choose. (Again, the lower resolution only affects what you see in the edit.)

Optimized Media (see below) does essentially the same thing. The main reason for using Proxy Media over Optimized Media is that Proxy Media creates files that can be shared with other editors (along with the project files). Proxy Media files are also smaller. And it may be that one or the other is faster to generate or plays back better on your computer—you'll have to run some tests. But the consensus is that Proxy Media is now the transcoding option of choice because of the speed of generating the files and improved playback performance.

As with Optimized Media, if you edit with proxies, at some point, you'll have to use the original footage for output. In some cases, that will mean a longish final render while the effects you added to the low res proxies get re-processed using the higher resolution original footage.

Typically, set *Proxy media resolution* to Half if you are shooting HD, to Quarter if you are shooting 4K, and One-Eighth if your camera has even higher resolution. Using DNxHR HQX (the default) is theoretically the correct choice for Windows, although ProRes would be the go-to on a Mac. However, that merely defines the codec: the file type will be the same as the original file itself.

You can generate Proxy Media for a clip by selecting a clip in a Bin, right-clicking, and choosing *Generate Proxy Media*. And like Optimized Media, you have to tell Resolve to use Proxy Media (under the Playback tab on the Edit page) in order to use the proxies you created. Under the *Playback* tab, select Proxy Handling. Here you get a choice of using proxies for the edit or not. The default is to use them.

And since there are no free lunches, generating proxies is also going to take time. It would be a good practice to take one representative clip, make a test proxy file, and see if you think the improvements are worth the cost before committing a whole project to a particular workflow.

When you generate Proxy Media via Resolve, the proxy file will automatically be linked to the original footage. If your camera creates its own proxies on the fly, you can also link the camera-generated proxies to your uncompressed camera footage instead of creating them in Resolve. That saves heaps of time.

Proxy Media and Optimized Media perform similar functions but not in the same way. Given the files you are working with and your computer's capabilities, it's possible that one will be better than the other. You should test both processes on your camera's footage and see which one, if any, substantially outperforms the other. I now use Proxy Media and hardly ever use Optimized Media anymore.

Optimized Media

Like Proxy Media, Optimized Media is useful when your camera's codec is a problem for Resolve to handle right out of the box. Here I'm talking about H.264/265 footage, but this really applies to any *interframe* codec. It's a common practice with hard-to-edit camera codecs to transcode the files via Optimized Media (or Proxy Media) using a codec designed for post-production such as DNxHR or ProRes.

In addition to the codec type, high resolution footage can also be hard for your computer to handle (more pixels per frame). If the footage is too high res for your computer to handle easily, you could also drop the resolution to less than the original when you optimize your files. Optimized Media files are, however, very large, and they are persistent (and must be cleaned out manually.)

You can create smaller Optimized Media files if you optimize using very high compression settings. But higher compression takes longer to process, and the resulting files are of lower quality (but that only affects what you see in the edit). In practice, optimizing to a different codec at reduced resolution while using a lighter (better quality) compression setting seems optimal (and is currently the default).

Without getting too much into the nitty-gritty, the H.264/265 codec family does not record video frame by frame. Instead, this type of codec records a frame (called an "I-frame") which is an actual frame, sort of like a jpeg. Then they break the next image up into, say, a checkerboard of 64 squares and look at the next few frames to see how much anything in these squares actually changes frame to frame (i.e., interframe). Then they record only those squares (called B-frames) that are actually changing—which is usually only a fraction of the whole video image.

This type of codec records only those squares that change for a few more frames until it has to send another complete I-frame. On the other end, some video app (such as Resolve) takes the I-frames and B-frames and makes whole frames from that data. This is why your computer needs so much processing power. It's got a lot of work to do just to turn the camera's video file fragments into complete frames that Resolve can play back and edit. If it didn't have to do this, playback would be a lot easier on your computer.

This is an over-simplified description, but it illustrates how H.264/265 and other in-camera *interframe* compression schemes don't record actual frames as such and why transcoding to a codec that does everything in honest-to-god frames improves playback performance. Resolve can usually re-create frame-by-frame video on the fly from H.264/265 footage, but that taxes your computer when it's also trying to render effects and do other things–hence the need to render and cache.

When you Optimize Media, Resolve takes the H.264/265 file and recreates a frame-by-frame video version. In fact, if you take a look at the Optimized Media folder, you'll see a bazillion .dvcc files—each one is a single frame of video. This is why Optimize Media takes a while to process and why the resulting files are so large. But once transcoded, these files are much easier to play back.

There are several settings to control how media gets optimized. These settings only matter if and when you actually generate Optimized Media. But set them even if you don't plan on using them right now. *Optimized Media Resolution* defaults to Automatically but can be set to the same resolution as your camera footage (Original) or to some lesser value. Typically, it's set lower—a lot lower. The choices are Original, Half, Quarter, One-Eighth, and a Sixteenth. For 4K, you might want to set the resolution to half or even one quarter. The image in the viewer may not look that good to you, but the final output using the original footage will be fine.

"Automatically" basically does the math for you by dropping Optimized Media's resolution from original to whatever the Timeline resolution is set for. If you don't need extreme settings, just use Automatically. Admittedly, the reduced resolution may not be pretty on your viewer while you edit, but it doesn't affect the final quality (as long as you don't output using reduced-resolution Optimized Media).

Setting the Optimized Media Format to a more highly-compressed version does make the files smaller, so use high compression settings if you don't have much room on your cache disc. Optimized Media's default codec is now DNxHR HQX. That's a high-quality compression setting. The resulting file size will be huge. You may want to lower the compression quality to reduce the file size (to DNxHR SQ or even LB). But with higher compression/lower quality settings, it may take a bit longer to transcode files.

If you plan to use Optimized Media, you must first generate it by selecting a clip or clips, right-clicking on one of the clips in a Bin, and selecting *Generate Optimized Media*. Once that's done, you will have to go to Edit page > Playback > *Use Optimized Media if Available*. Footage can be transcoded in the Media Pool or in the Timeline itself (right-click on the selected clip). You can optimize a single clip or batches of clips. However, if the clip is long and you only intend to use a small portion, it's better to rough cut the clip in the Timeline and optimize just the portion you actually intend to use (or create a subclip in the Bin and just optimize that).

Optimizing takes time—lots of it. Shoot a test clip and see how much Optimized Media actually helps. I have one camera that shoots AVCHD, and Optimized Media substantially improves Resolve's handling of that footage. Another camera records in XAVC S and Optimized Media appears to deliver less improvement.

> **CAVEAT**: Optimized Media files are massive and should be deleted when the project wraps. BMD now says, "Deleting optimized media from DaVinci Resolve now must take place *manually* by removing the files in the CacheClip/Optimized Media folder in your OS." However, that button may be back by the time you read this.

Render Cache

All of the above settings are designed to take some of the load off your CPU/GPU so that your computer will likely play the Timeline smoothly without the need to halt and render as often. I'm guessing that at some point, you will nonetheless come up with some effects combo that can only play in real time on a smoking hot box (which you probably don't have). In that case, your machine will have to halt and render, so how you set up Resolve to handle the inevitable rendering and caching is important. In that regard, Render Cache is first among equals.

Generally speaking, Render Cache (Smart/User) controls the rendering of effects added via the Edit page. Fusion Output Render Cache controls the rendering of everything *upstream* of the Fusion page (including Fusion), while Color Output Render Cache controls the rendering of everything *upstream* of the Color page (including Color), which at that point is really everything. But even so, Fusion and Color page rendering is enabled/disabled depending on the higher-level Render Cache setting.

Set Project Settings > Master Settings > Optimized Media and Render Cache > *Render Cache format* to DNx *something*. DNxHR (an Avid codec) is one Resolve particularly likes (use ProRes if you're on a Mac). You can't change Render Cache's resolution because that's determined by the Timeline. The only thing you can change is the codec and compression level.

As with the Optimized Media and Proxy Media settings, higher compression results in lower visual quality, but only you see the reduced visual quality (unless you are outputting using these files). And more compression equals smaller files but may take longer to create. You'll have to test and evaluate the effect of these settings on your eyes and on your patience. The default is probably fine.

BMD's official DaVinci Resolve Reference Manual says that "choosing a more highly compressed cache format makes real-time playback possible on less capable computers with slower and smaller storage, at the expense of slightly compromised image quality." That's the theory, anyway. However, I couldn't achieve better FPS or significantly faster renders with the compression set to even the lowest quality setting. Your mileage may vary.

However, higher compression does equal smaller files. And smaller files are faster to move around. But higher compression also requires more calculations and takes longer. Higher compression could be useful on smaller, slower drives. But you should be storing these files on large, fast drives anyway.

The default is DNxHR SQ (or ProRes LT on a Mac) which is probably right. If you are shooting HDR or certain other high-end formats or want to preserve the Alpha channel, you'll need to render using 16-bit float uncompressed such as DNxHR 444 or ProRes 4444.

Like Optimized Media, you have the option of using the Render Cache files instead of the original footage when you do the final render for output. That can substantially speed up the final render of long-form projects because a lot of the work has already been done. But if you plan to do that, you must set the *Render Cache Format's* compression scheme appropriate to your original footage. Otherwise, Resolve will create the final render using highly compressed "Optimized Media" files, which won't look very good when used as source (see below). Here are the suggested settings:

DNxHR 444-HDR for s-log footage
DNxHR HQX for 4K
DNxHR HQ for HD

Set *Enable background caching after 5 seconds* to either one second if you want it to start rendering as soon as you complete an edit (and are AFK) or to 30 seconds if you want Resolve to make sure you are finished with what you're doing before it takes over the GPU. The default setting of five seconds is probably fine.

It's probably a good idea to leave *Automatically cache transitions in User Mode* on and turn the other two automatic cache operations off. The latter two cache systems are highly processor-intensive, and on most machines, you'll probably want to enable those manually as needed.

Background Caching

If the check box is unchecked, *background caching will not occur*—no matter what the rest of the settings are anywhere in the program.

That can be useful if you don't want Resolve to halt and render whenever it's idle as it sometimes wants to do. I've found that, in practice, it can sometimes be useful to turn background caching off so that I have complete manual control over the rendering process and can edit without constantly fighting the machine for resources. (And I would be remiss if I didn't point out that although BMD calls this "background" caching, it actually occurs in foreground when nothing else is going on.)

Working Folders

Where the Proxy Media files, Render Cache, and Gallery stills are stored is initially determined in Preferences. You can change those locations for a particular project. You can also discover where Proxy Media files are stored on your computer. These files are portable, and you could transfer a project to another machine along with the Proxy Media and let someone else use the Proxy Media files while editing on another computer without having to regenerate them.

Using Optimized Media, Proxy Media, or Render Cache for Final Output

At the time of final render, one of the options (under Custom) is to Use Optimized or Proxy Media for your final output instead of your camera-original footage. You could also use Render Cache files for this purpose as well (but don't).

It is usually *not desirable* to use Optimized Media, Proxy Media, or the Render Cache for your final output. To make that work, you would have had to create those files using high-resolution/low compression settings (which, to my mind, negates the whole concept of optimization). However, using these files would make the final render a lot faster (because of the pre-rendering), and that may be important on long-form projects. But that's not the default, so if you do nothing, Resolve will always use your camera original and not Optimized Media, Proxy Media, or the Render Cache files for final rendering.

Timeline Proxy Resolution

Once all of those settings have been accepted or changed, the next place to look at is the *Timeline Proxy Resolution* (which is in the Playback pull-down menu on the Edit page). All Timeline Proxy Resolution does is reduce the resolution of your Timeline Viewer. It requires no processing or pre-rendering and can be turned on and off on the fly. You can choose to work with a Timeline with less resolution than the original footage to reduce the need to render as often.

Timeline Proxy Resolution has no effect on the quality of the final render or quality of effects and grades applied in edit (except for what you see while editing). Operating with reduced resolution means that the Timeline will playback more smoothly, more often, and, when Resolve does need to halt and render, creating a Viewer file that will play smoothly won't take as long.

While there are many different rendering, optimizing, and caching scenarios, if the Timeline playback glitches or drops frames, try lowering the Timeline Proxy Resolution first. Timeline Proxy Resolution is instant, and you can change it at any time without having to re-render everything. Timeline Proxy Resolution is basically a preset that allows you to cut the Viewers' resolution with a single click. Go to Edit page > Playback and set Timeline Proxy Resolution to half resolution or lower. If you need to see your work in full res, simply turn Timeline Proxy Resolution to *Full*.

While Timeline Proxy Resolution reduces the computational load to a degree, it really only affects the calculations needed to supply the Timeline Viewer and does nothing to reduce other calculations such as those needed to process Timeline transitions and effects.

Show All Video Frames (Not)

If your camera's footage still does not play smoothly on your PC, even with Timeline Proxy Resolution set to half or quarter, look at the Edit page's Viewer options (under the three dots at the top right of the Viewer). You have a choice to either *Show All Video Frames* or not.

If it's checked, Resolve will try to process all of the frames on the fly, and the audio can be glitchy (if playback is less than the Timeline's actual FPS). If it's off, Resolve will drop video frames if necessary so that the audio track plays more or less smoothly.

If your computer can handle the playback while rendering, this setting does nothing. It simply prioritizes how Resolve is going to degrade if and when your computer can't keep up. And the video glitches and stuttering will get taken care of at the time of final render (fingers crossed). To check or not to check is a personal preference. Personally, I prefer smooth audio, so I uncheck this box.

Timeline Resolution

The third option to improve playback performance is to reduce the actual resolution of the Timeline itself while you are editing (Project Settings). In that case, playback and rendering will take place at this lower resolution during the edit. If you shot in 4K but set your Timeline Resolution to 1080p or even 720p, the Timeline will play smoothly more often. If you are going to output in 1080p or 720p, you should probably operate this way anyway.

Editing 4K footage in a 1080p Timeline, then changing the Timeline resolution back to 4K for output will reduce playback/render issues during the edit. This is, in fact, a very common workflow practice. However, when you change the Timeline resolution, *everything will have to be re-rendered* in the higher resolution. On the other hand, knowing that in advance lets you plan for it. Many editors would rather have that one long render at a time and place of their choosing than the constant little interruptions.

However, while editing in high resolution can create bottlenecks, by far the largest drag on your computer is the choice of codec (and depending on the camera, you may not have much choice). Some camera codecs require a lot more CPU/GPU activity to play than others. Reducing resolution is an easy way to gain performance but doesn't directly address the main reason Resolve is struggling on your machine. To truly fix that, you'll need to transcode your camera's footage to a codec that's easier for Resolve and your computer to play back.

What's the Cache?

I admit the above heading is a cheap pun, but it really does apply to Resolve. It's no joke—there are so many different rendering/caching subsystems in Resolve with so many different ways to activate them, it's hard to keep track. Of course, BMD could have just implemented one render cache like the other guys, but having multiple caching systems—essentially a separate one for every major functional area—is a much more powerful toolset than putting everything in one basket.

But this profusion of render subsystems means that there are more things to consider when you're first setting up Resolve and more things to fiddle with when you're working with Resolve. But that is also what allows Resolve to implement powerful features without you having to have a Pixar-level machine to run it on. And the controls for all these rendering and caching subsystems are mostly "off/on/auto" type inputs anyway—you just need to know where the buttons are and what they turn on and off. It may not matter to you right now, but the ability to render certain effects while suppressing rendering for others can be a real-timesaver if you are working on a complex project with lots of effects.

Rendering Process

Rendering is just another word for "processing." Camera footage has to be processed in order for it to be imported/ingested. If the footage is not recorded in actual frames (H.264/265), Resolve will have to decode the file and recreate it in frames because editing is based on frames. Resolve also generates a database file to track what's in the Timeline and what's been changed. And it needs to generate thumbnails for the Timeline as well as a proxy video file for the Viewer plus audio files, waveforms—lots of computations and file updates. That's a lot of processing, particularly if you are trying to do it in real-time.

Your camera doesn't create these files: Resolve does. So the render process really starts when you first ingest media. Resolve has to unpack your H.264, BRAW, .mov, or other file formats and process them to create the Timeline, the thumbnails, and the Viewer's video and audio files. Fortunately, this happens automatically and requires no input from you (usually).

Once a clip is in the Timeline, it typically plays just fine on virtually any recent vintage PC or Mac. If Resolve does have trouble with your camera's codec, you could always change it to something Resolve finds easier to handle via Optimized Media, Proxy Media, or Media Management, or in some cases, in the camera itself.

When you make an edit, add a transition, or apply an effect, Resolve will have to re-render the Timeline's internal files, update the thumbnails, and modify the video and audio files used by the Viewer during playback. Your machine may be powerful enough to perform all these changes while playing the clip, in real-time, without having to halt and render them out. You may get a glitch here and there, but most transitions and many effects process on the fly during playback on reasonably powerful machines.

When Resolve halts and renders, it caches the results of the render in the Render Cache, which is the first folder/drive on the list in Preferences > Media Storage Locations. Once you've rendered a clip, playback will be from the cached version of the rendered Timeline, and it will play at the correct FPS (until you change something).

However, Resolve does just enough render calculations to enable the clip to play smoothly. It does not render the clip and effects down to the level of original footage until the final render at Delivery. This behavior greatly improves render performance. Resolve needs to do only enough math during rendering so that your computer can handle the rest during playback. Resolve's AI is not perfect, and sometimes even clips that have been rendered can stutter a little (particularly if your OS is doing something in background while you are playing).

In some cases, rendering is such a processing burden on your GPU/CPU that the Timeline can't play smoothly at the proper FPS, or it won't play at all. In that case, Resolve will need to halt and render. When it does that, you'll see a red line over the portion of the clip that requires rendering, and it'll change to blue as the render progresses. In such cases, Resolve is going to store or "cache" the resulting rendered clip so that it won't have to be done again (until you change something).

There's no way to avoid rendering. But rendering that occurs while you play the Timeline is invisible. Having to wait for Resolve to halt and render a change is more than visible—it can be annoying. Almost all of the previous settings were chosen to reduce how often Resolve needs to halt, render, and cache in order to play the Timeline smoothly. But since halting to render is still going to happen some of the time, how you use Resolve's rendering systems will have a huge impact on your workflow.

Resolve slices and dices the rendering tasks across several subsystems in order to give users maximum flexibility and control, but all that power comes at a price: *complexity*.

There are at least five render behaviors in Resolve:

(1) Resolve can render on the fly—during playback, (this happens more often than you'd think). Rendering has to be done every time the clip is played. If your computer is under-endowed, playback may occasionally stutter or glitch. In some cases, the need to render is so overwhelming that the Timeline will not play back at all until the clip has been rendered and cached.

(2) Playback > Render Cache > *Smart mode* can automatically identify a clip that Resolve *thinks* needs to be rendered in order to play smoothly, and it will render it when you are AFK (or have not moved the mouse for five seconds or so). It also caches the results in the Render Cache so the clip doesn't have to be continually re-rendered (until you change something). If Smart mode fails to recognize that a clip can't be played smoothly, you can manually "flag" the clip for rendering (see below). Occasionally, for one reason or another, rendering may not commence in Smart mode even when a clip is manually flagged. In that case, Smart mode thinks the clip should play smoothly, so it doesn't automatically render it. Sometimes Smart mode gets it wrong, and the clip does need to be rendered. In that case, simply switching to User mode will usually trigger the render.

(3) Playback > Render Cache > *User mode* lets you manually initiate rendering. When User mode is enabled, you can initiate a render by right-clicking on a clip and checking *Render Cache Color Output* whether Resolve thinks it needs to be rendered or not. This step is known as "flagging" the clip for rendering. In User mode, *you* get to decide if the playback is acceptable or if you want to stop and render the clip. In some cases, it's obvious to Resolve what you want to do, and simply being in User mode will initiate renders more often than Smart mode. But if you are in User mode and need to render a clip, manually flag it to force a render.

(4) In addition to caching at the Timeline or clip level, you can also cache individual Nodes on the Fusion and Color pages.

(5) And of course, off (None) is always an option.

Operational settings for all of these Render Cache behaviors are covered below. Furthermore, Casey Faris has an excellent render settings video on YouTube that's based on hundreds of *actual tests* he conducted. I strongly recommend you watch it.

Render in Place

If this embarrassment of riches is not sufficient, starting with Resolve 17, there's yet another render system. And this one packs a wallop. *Render in Place* lets you render a clip on the Edit page Timeline.

Render in Place creates what is essentially a composite clip that incorporates all of the effects that have been applied to it—OpenFX, OFX, Fusion, and even grading. Rendered. Cached. Done.

To render a clip in place, right-click on it and select *Render in Place*. You get to choose the Format, Codec, and Type. In this instance, Type includes resolution and bit depth.

Since you can choose the format, codec, and resolution for Render in Place, you could say it's essentially Optimized Media/Proxy Media/Render Cache/Color Node Cache/Fusion Cache all rolled into one. You can also decompose the composite Render in Place clip back to its original form should you need to change anything. (Of course, it'll have to be re-rendered.) And you can group clips together and render them as a group.

Deep Dive

If you want to delve even deeper, see the chapter on Render Systems.

Suggested
DaVinci Resolve Settings

**DaVinci Resolve > Preferences > System >
Memory and GPU > GPU Configuration > GPU processing mode:**

- *CUDA* for Nvidia cards or *OpenCL* for AMD
- GPU selection Auto (usually—see text)

**DaVinci Resolve > Preferences > System > Media Storage >
Media Storage Locations**

- The first location on the list should be your fastest drive

DaVinci Resolve > Preferences > System > Decode Options

- Decode H.264/265 using hardware acceleration
- Select AMD if you have an AMD graphics card

**DaVinci Resolve > Preferences > User > UI Settings >
Workspace Options**

- Reload last working project when logging in

**DaVinci Resolve > Preferences > User > UI Settings >
Project Save and Load > Save Settings**

- Live Save
- Project Backups

DaVinci Resolve > Preferences > User > Playback Settings

- *Hide UI Overlays*
- *Minimize interface updates during playback*
- *Performance Mode* (Automatic)

File > Project Settings > Master Settings > Timeline Format

- Timeline resolution (usually same as camera footage or less)
- Timeline frame rate (usually same as camera footage)
- Video bit depth (8 bit)
- Proxy Media resolution (see text—typically one half or one quarter)
- Proxy Media format (DNxHR HQ/ProRes 422 Proxy)
- Optimized Media resolution (see text—typically one half or one quarter)
- Optimized Media format (DNxHR SQ/ProRes 422 Proxy)
- Render Cache Format (DNxHR SQ/ProRes 422 Proxy)
- Enable background caching after 5 seconds
- *Save your changes*

Edit page > Playback > Timeline Proxy Mode

- Half Resolution

Edit page > Playback > Render Cache

- Smart

Edit page > Viewer Options (three dots at top right of viewer)

- *Show All Video Frames* (uncheck)

Exercises

Exercise Uno

There are a lot of ways to learn DaVinci Resolve 18. It is my firm belief that instead of being shown a lot of fancy features which you may never use, a better way is to start with one very simple yet practical end-to-end project and from there progress to more complex ones. Along the way, you'll learn Resolve well enough that you'll be able to figure out its more sophisticated features. (And the more-advanced YouTube tutorials will be easier to follow.)

If you are somewhat familiar with Resolve already, this first exercise may seem over-simplified. It is designed to expose new users to the main controls and basic operations. I think most people's frustration with Resolve is simply because they can't find the controls they are looking for and don't fully understand what they do. IMO creating a simple project from end to end is a better approach than simply demonstrating all the whiz-bang features and hoping that the user can figure out how to use them (and remember where they are).

> **CAVEAT:** The folks at BMD must stay up late at night making changes and improvements and applying bug fixes to Resolve. While every keystroke and mouse click in the following exercises were checked numerous times, with every update something somewhere moved or simply didn't work exactly like it did in the previous release. At the time of printing, it was accurate (Resolve 18.0), but depending on what version and build you have or may have, later on you may find some directions no longer work exactly as described (or, more likely, they have been renamed). Patience and perseverance are the only workarounds for that.

For the first exercise, you'll need to shoot a short video clip using your camera set to the resolution, frame rate, codec, and format you intend to use for later projects. Nothing fancy, just a simple talking head, a few minutes long, shot "run and gun." Start and stop once or twice. Flub a line. Bump the mic. Get the kind of footage most people usually wind up with when shooting the ubiquitous talking head scenario.

If you already know the basics, you can start a new project, import your video, and skip to The Actual Edit (below).

In this exercise, you are going to import footage into Resolve (probably as two or three clips), edit out the bad parts, fix the audio track, and export an output file suitable for YouTube, Vimeo, or Twitter. Pretty standard stuff. But even so, it'll give you the opportunity to learn the Resolve User Interface (UI) and explore its essential functions.

For the audio, record two mono tracks. It's a good practice to put the main mic on one track (wireless, desk, or boom) and use the second track for safety—a shotgun perhaps, or just Track 1 duplicated at a reduced (50%) level. For a talking head, aim the mic at the talent's throat/sternum, not directly at the mouth, so you pick up the chest resonance (and miss the plosives, so you'll have fewer problems to fix). If your mic allows, turn the high pass/low cut filter on to reduce rumble (better to not have it in than have to take it out later). Get the mic fairly close to the talent (a foot or so) to pick up less of the room's coloration. And do what you can to reduce ambient (room) noise.

If it's not possible for you to shoot your own sample footage, BMD has sample footage and projects on their website you can download.

BLACKMAGICDESIGN.COM/PRODUCTS/DAVINCIRESOLVE/TRAINING

However, I believe that you'll get more out of these lessons if you think of some project you might conceivably create (unboxing video, cooking demonstration, home repair tip) and shoot a few short video clips with your own camera.

IMPORTANT: At the end of the day, you need to be able to create a finished, distributable video. This means you have to know in advance what the end game is. So start with a script or at least an outline of what you want to end up with and work from that. Or, better yet, storyboard (on a whiteboard) each shot/scene.) You need to have a plan, visually mapping how the video is supposed to turn out. Simply learning how to splice two clips together and add an effect won't teach you much about the editing process. Furthermore, I would suggest your project should include some talking head, voiceover (audio track), stock footage, stills, and text/titles because most finished videos tend to have all or nearly all of these elements.

If you want to learn how to do Chroma Key or grade a shot, it needs to be footage shot with *your* green screen, *your* lighting, and *your* camera. Chroma Keying professional green screen footage shot with perfect lighting isn't all that hard and teaches very little. Likewise with grading. Many of the practice videos I've seen use stunning footage and were shot on very high-end equipment. That's not likely to be the case for someone just learning Resolve. All you need is just a few minutes of footage anyway. And shooting your own does force you to look at your camera's manual and decide what format, frame rate, and codec is best for your workflow and desired result.

Storage Locations

Per previous instructions, using your computer's OS, create a *cache folder* on the fastest drive you own and make sure that folder is the *first item on the list* in DaVinci Resolve > Preferences > Media Storage. Create a *media folder*, and put the clips you shot (and any other media) there. Also, create an *output folder*. Put these on the list, too, just not first.

Launch Resolve

Go to Project Manager and create a New Project. Call it "Exercise Uno 01" if you want. Make sure the Preferences and Project Settings are to your liking (using information from the previous chapter).

Head to the Media page. Go to Workspace and *Reset UI Layout*, so we're all on the same page—literally. You can import clips while you're on the Edit page, but the Media page is made for it, so we're starting there.

Cleanup on Aisle 16

On the Media page, there is a 16-track mixer (*Audio*) that you don't need at this point (and never will) and a *Metadata* panel you also don't need right now. Turn those off to clean up the screen so you can make the Source Viewer larger. You can shrink the Media Pool/Master Bin, too. While Resolve's UI customization is more limited than Premier or Final Cut, you'll have enough screen space if you take the time to kill those parts you don't need or as you finish with them.

Importing Media

At the top left of the screen, Media Storage will be highlighted. Immediately below is a list of storage locations. Double click on the media folder where your clips are stored.

Scrubber

Pre-Editing

You could just select the clips you need and drag them from Media Storage to the Master Bin. But it's a common workflow practice to pre-edit long clips by double-clicking on the clip, then playing/scrubbing it in the Source Viewer and selecting in and out points (I&O) so you grab only the footage you actually intend to use. Do this by grabbing the Viewer image and dragging it to the bin, leaving the dross behind. (The I&O dots are adjustable, and Alt+X clears the I&O points.) There are many different ways to view, preview, scrub, play, label, color code, and store media, and you should adopt a consistent way of doing that (workflow).

Viewer Transport Controls

The Up/Down/Left/Right arrow keys let you navigate around in the Master Bin. Normally, these keys are attached to the selected Viewer, where they control the transport (Up = start and Down = end). Unfortunately, there's no visual indicator to tell which function has been selected, but it's easy enough to figure out. Simply click on Media Storage, the Master Bin, or the Viewer, and the arrow keys will follow your selection.

Click on a clip in Media Storage, and it'll pop up in the Source Viewer so you can play it. Play the clip using the space bar to start and stop. Or scrub it in the Viewer using the scrubber for more precision. Having metadata up provides useful technical details about the clip. If you just want a quick view, the thumbnails in Media Storage are live–just scrub over them.

If you don't have a Blackmagic hardware control panel, you can use the J-K-L keys to review footage. J = play backwards, L = play forward, and K = stop. Holding down K and pressing J or L lets you play forward or reverse in slow motion. You can also hold down K and quickly pop J or L to move forward or backward one frame at a time.

The transport controls at the bottom of the Source Viewer are typical (Head, Play Reverse, Stop, Play Forward, Tail, and Loop). Play the clip (or scrub it) and find the section you want. Mark it with the I&O keys.

Drag the clip to the Master Bin. If you selected I&O points, Resolve will create a "Subclip" bounded by the I&O points. You could also import an entire media folder into the Master Bin. When you have imported all the clips you need, kill Media Storage.

> **CAVEAT**: When you drag your first clip to the Media Pool/Master Bin, if its frame rate differs from the frame rate in the Project Settings, you'll get a warning message. And you'll have to either change the project's frame rate to match the clip or vice versa. *The Timeline's frame rate cannot be changed* once a file has been added to the Media Pool (which also creates the Timeline).

Source Viewer

While we're still on the Source Viewer, let me point out a few things. At the top left, you can select the frame fill percent (defaults to Fit). That doesn't change the size of the Viewer, only the size of the image in the Viewer. To change the size of the Viewer, look around for panels you no longer need and deactivate, then grab the line immediately below the Transport controls and adjust.

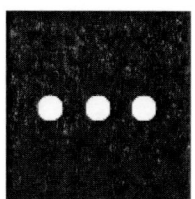

At the top right on the Source Viewer are three dots. Each page in Resolve and many of the control panels have a "three dots" menu, and there's usually something interesting (and poorly documented) behind the three dots. Unselect *Show All Video Frames* if you haven't done so already. And, if you haven't done so, set Playback > Timeline Proxy Resolution > *Half Resolution* and Render Cache to *Smart*. Viewing the edit in reduced resolution may look shabby, but it will speed up rendering and improve playback. And it doesn't affect the quality of the final output.

I&O Points

If you had one really long take, you could pull four or five SOT (sound on tape) subclips from it for easier editing. As an alternative, you could switch to the Edit page and *drag the entire clip from the Source Viewer directly to the Timeline,* and then cut it up. In Resolve, there are almost always two or three ways of doing anything if you know how.

> **CAVEAT**: When you select I&O points in the Source Viewer, what you do next determines if you will be able to easily increase the length/duration of the clip once it's in the Timeline.
>
> If you set I&O points and drag the selection directly *into the Timeline*, you will be able to increase the clip's length/duration in the Timeline (using the Resize tool).
>
> However, if you set I&O points and drag the subclip *into a Bin*, you will have created a separate *subclip* that's bounded by the I&O points. It's easy to shorten a subclip, but it's hard to lengthen it. If you need to create separate subclips and keep them in Bins, be generous when setting the I&O points (giving yourself several extra seconds on each end should you later decide you need a dissolve).
>
> In the Master Bin, if you view media as thumbnails, you probably won't be able to see the appended "subclip" label. For this reason, I find the Metadata View easier to use. (List View works as well.) Also, note that the names of additional subclips you create do not get incremented. You'll need to go into the Master Bin and rename them if you create more than one from a single clip (if that matters to you).

Master Bin

Once a clip is in the Master Bin, *where you stop scrubbing is going to be the thumbnail image.* Otherwise, it's the first frame of the video, and that may not be representative (but you can change that by scrubbing the thumbnail to the frame you want and carefully easing your way out).

If you want to create multiple Bins for a large project, *from the Master Bin*, select a bunch of clips, right-click on them, and select *Create Bin With Selected Clips*. It will show up in the Master Bin. You can rename it to whatever you want. Or, in the Master Bin, just right-click and *New Bin*, then drag clips to it.

At the top right of every Bin is a slider that lets you change the size of the thumbnails. Next to that are three icons that let you display clips either as thumbnails (with or without metadata) or as a list. You can Search by clip name (or some other field) and use the down arrow icon to sort clips by some attribute like type, reel name, date created, etc.

Now move to the Edit page. Drag the clip with the black and white reference to the Timeline first. (You did shoot a B&W reference, didn't you?) Drag the remaining clips/subclips into the Timeline roughly in the order you intend to use them. If necessary, adjust the -/+ slider (mid-screen right) to change the temporal view (time base) of the Timeline. (The Alt key and mouse wheel works nicely also.) Early on, you'll probably operate with large thumbnails and a small Source Viewer. Later, you'll want the reverse.

Setting White and Black Levels

Select the White/Black reference clip and put the play head over the shot of the reference box (described in an earlier chapter). Move to the Color page. Turn Gallery off.

Go to Workspace > Video Scopes > On (Ctrl+shift+W). Click the single-view scope for now (top right). For what follows, you'll want *Parade*. Also, click on the "X" (top left), and the Scopes will find a nice home.

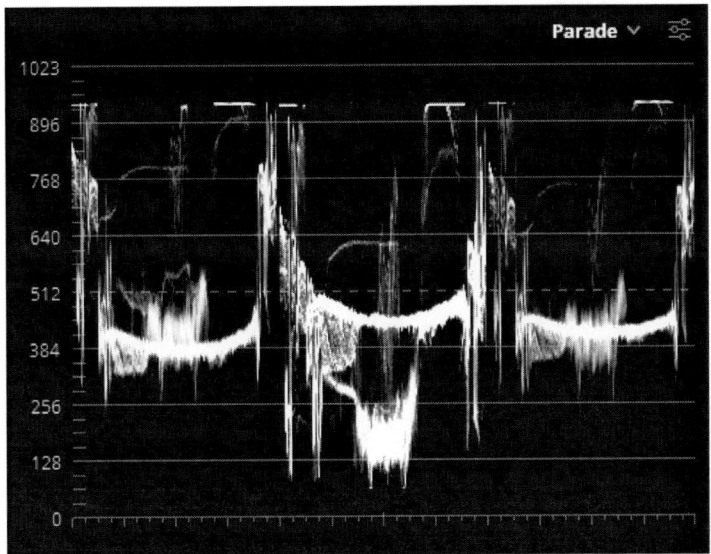

The Parade scope shows separate RGB channels

You should be able to tell, spatially, where the white and the black reference lines are located on the Parade scope. Adjust Gain's *Luma* control (horizontal wheel) such that the white chip is around 896 (if it is truly ultra white), and there are no "fuzzies" or anything else hitting 1023. Then adjust Lift's *Luma* control such that the black reference line is ever so slightly above 0. This procedure assumes the reference was shot in the same lighting as the talent.

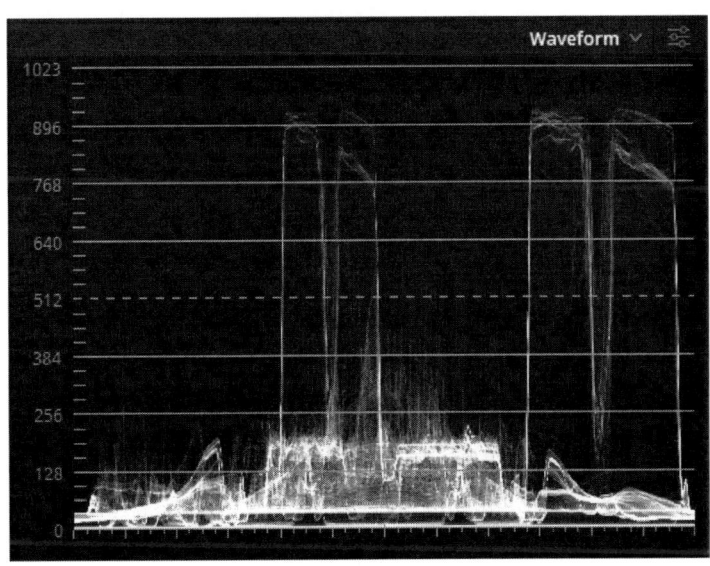

Waveform combines RGB into a single Luma display

White Balancing

Change the scope from Parade to Waveform. Although I'm sure you white balanced your camera, I'm also fairly confident the white chip reference appears as slightly separate RGB lines and not as a single, solid white line. Even if the white chip is not actually 100% white, your camera should have made it so. In my experience, camera white balancing systems are not that accurate and can stand to be tweaked a little.

WB Eyedropper

The eyedropper (under Primaries) is for white balancing. Click on it (it's sticky), and apply it to the white reference chip. Look at the effect on the RGB lines on the Waveform scope. I'm sure that helped merge them a bit. You could also tweak White Balance by switching from Wheels to Bars and adjusting Gain's R, G, or B bars to create (more or less) a single white reference line (so that the R, G, B lines merge to create pure white), but you really do not have to get these perfectly aligned.

You may find that adjusting one color affects the other colors, and not in a good way. Find the *L. Mix* control below Offset. The default is 100%, which means Resolve will try to "help" you when you tweak one color by adjusting the other colors so that Luma levels remain constant. Normally, this is a good thing. If it's interfering with your color corrections, you can dial *L. Mix* down (to 0 if necessary) to uncouple Chroma changes from Luma.

Once you've achieved white balance, adjust Gain's Luma control again if necessary so that the white line is still about 896 or so (depending on how white the reference is), and there are no RGB "fuzzies" hitting 1023 or anything else.

Another option is *Auto Balance* (the "A" next to the White Balance eyedropper). Auto Balance will automatically white balance a clip (if there's a white reference in it).

The thumbnail at the top right is one of those Nodes that Resolve is famous for (and rightly so). The Luma and WB corrections you just made were incorporated into the default Node. Right-click on it and name it WB/LUMA. Notice that there are two or three tiny icons (bars and curves, perhaps the eyedropper) at the bottom to indicate what that Node is doing.

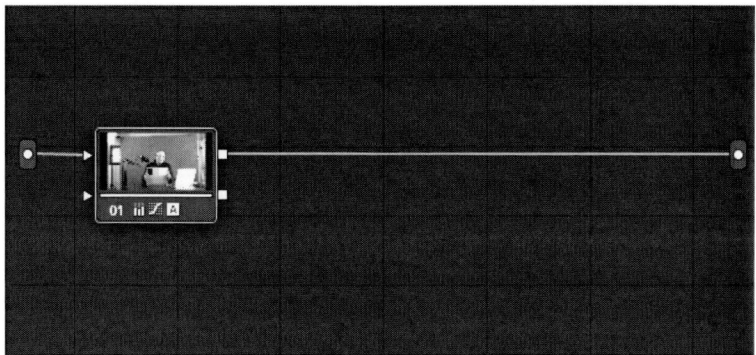
WB/Luma Node

Copying Nodes

You can copy these basic Luma and white balance corrections from the WB/LUMA Node to other clips in the Timeline (assuming that they were shot with the same lighting and exposure).

Still on the Color page, at the top right of screen make sure Clips is on. Select the thumbnail of the clip with the WB/LUMA Node — this is the one you want to copy FROM—and Ctrl+C. Select the clip you want to copy TO and Ctrl+V. The attributes of the WB/LUMA Node have now been copied to the selected clip. Repeat for any other clips shot with the same lighting and exposure. This helps achieve consistency from clip to clip (assuming they are similar).

To remove these attributes, click on a clip, then right-click on its WB/LUMA Node and select *Reset Node Grade*.

To copy a Node's attributes to more than one clip at a time, select the thumbnails of the clips you want to copy TO using shift-click to select multiple clips. Then place the cursor over the clip with the WB/LUMA node you want to *copy* FROM, and press the *middle mouse button*. Now all the selected (target) clips will have the same basic WB/LUMA Node attributes as the reference clip. (And you can use this technique to duplicate whatever you've done to any Node.)

Edit Page

Return to the Edit page. You are now done using the reference clip, so delete it. This leaves a blank space in the Timeline. Click on the blank space, then right-click and select Ripple Delete. (Or just click on the blank space and press the Delete key.)

Click on the Mixer (top right) to activate it. This time it's a more useful stereo fader. In the Timeline, right-click on Audio 2 and select *Delete Empty Tracks* to clean up the screen. Grab the bottom of the Audio 1 panel if you want to adjust its size. Ditto for the very top line of the Timeline. Deleting unused tracks and shrinking the Timeline will expand the Viewers if you need to.

Press the *Home* key to force the play head to the beginning of the Timeline (fn+left arrow on a Mac).

Now play the first clip in the Timeline. If the clip is a talking head recorded with dual mono tracks, one of the tracks will likely be preferable over the other. Stop playback, right-click on the audio track of the clip and select Clip Attributes > *Audio*. Select Stereo, and pick one of the two tracks so that both Source Channels use the better of the two. (That will usually be Track 1 if you recorded a safety on Track 2.)

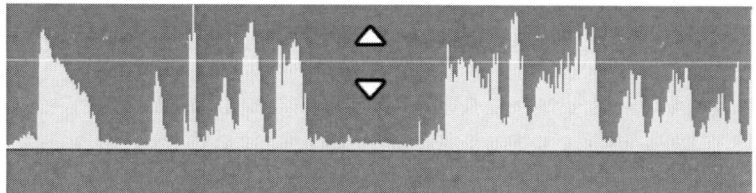

In-Track Volume Control

Notice that when the track plays, the Mixer has green, yellow, and red pips on each track's audio display. These pips are persistent (for a short while) and display the peaks. With the Mixer's faders in the default position (-10 dB), play the track and monitor the mixer's peaks. If you see a lot of red, stop.

There is a thin white (hard to see) horizontal line in the Timeline's audio track. That's the track's volume control. Up and down wedges will appear when you hover over that line. Adjust in-track volume control so that peaks are in the yellow but only occasionally and never in the red (we'll fine-tune the audio later).

You can only adjust the in-track volume control when the track is stopped, so you may have to go back a forth a few times, playing, stopping, and visually adjusting to get the level where you want it. Basically, you don't want to see any red. Usually, you can do these (rough) adjustments by eye simply by looking at the in-track audio waveform and adjusting the in-track audio level. You could also right-click on a clip and select Normalize Audio Levels which limits any peaks to the level you pick (-9 dB is the default).

Repeat setting the Source Channel audio for the remaining clips.

If the clips are from the same audio recording setup, you can copy the volume settings from the first clip to the rest. Select the source clip and Ctrl+C to copy the volume attributes from the adjusted clip. Select the target clip(s) and **Alt+V** to paste them to the other clips. Note this is **Alt+V** not Ctrl+V. Select Audio Attributes > *Volume*. *To see what attributes have been set for a clip, right click on a clip and select Clip Attributes. Note that when pasting attributes you do have some choice of what attributes to paste and which to ignore. However, you'll have to change the Source Channel track selection for each clip manually (don't ask).*

Most of the time, I just pull the entire clip into the Timeline, set the Luma, Gain, and Lift and white balance, select source tracks, adjust basic audio levels, *then cut the clip into pieces*—that negates the need to copy various attributes across multiple clips and ensures that all of these corrections are identical.

With the audio roughly set, reduce the audio track size to normal (or even less).

FPS Meter

Frames Per Second

As described earlier, when the Timeline is playing, the Frames Per Second (FPS) meter will pop up at the top left of the Timeline Viewer. A green dot means that the playback speed is equal to the actual frame rate of the Timeline as set in Project Settings—no stuttering (not very often anyway). A red dot (an FPS lower than the Timeline's actual setting) is problematic. If you are having any issues playing the Timeline smoothly without glitches, refer to the chapter on Resolving Performance Issues.

While you are still on the Edit page, you can grab some of the panels you don't need and reduce their size, which increases the size of the Viewers (there are limits, but do what you can). For example, when you have dragged all your media into the Timeline, you can kill the Media Pool. Also, there is a single rectangle at the top right of the Edit page (if the Metadata panel is off) that controls how many Viewers are active. Click on it to kill the Source Viewer when you no longer need it.

Color Code

If you right-click on a clip and select *Clip Color*, you can color-code the clips in the Timeline. A good practice is to color-code all of the SOT interviews one color, standups and talking heads another color, additional colors for B-roll, drone footage, stock footage, Optimized or Proxy Media—whatever. That will help you visually ID various types of clips in the Timeline.

Control+S to Save.

The Actual Edit

If you just joined us, we're on the Edit page. Press the *Home* key to move the play head to the beginning of the first clip in the Timeline. If you're used to working with Premier or Final Cut, you'll find that the Home/End and arrow keys operate much the same:

> **UP** = nearest break/edit on the left
>
> **DOWN** = nearest break/edit on the right
>
> **LEFT** = move left one frame at a time (plays backward when held)
>
> **RIGHT** = move right one frame at a time (plays forward when held)

Select and play the first clip (space bar). Stop (space bar) when you get about to the point where you want the video to start. Scrub to the exact frame using the Timeline Viewer's scrubber, the J-K-L keys (to play forward/backward or stop), the arrow keys, or just grab the play head and move it around using the mouse. Adjust the temporal slider (or use one of the three zoom presets adjacent to the slider) to zoom in or out as needed.

A handy feature (that should be the default IMO) is having Resolve automatically select the clip you have under the play head as the target for whatever comes next. To do that, go to the top of the page and select Timeline > *Selection Follows Playhead*. Now, the clip under the play head will be automatically selected for whatever operation comes next. Setting up the Timeline this way eliminates a lot of mouse clicks.

Press B (razor blade) and use it to cut the clip. Since the video and audio tracks are linked at this point, cutting either will cut both. Then press A (cursor), select the footage to the left of the cut (if not selected automatically by the play head), and press Delete. The selected clip will be deleted, leaving a hole. You can proceed with other edits, but if you want to close things up, click in the area where the clip used to be. That area will turn light gray. You can either press the Delete key or right click and select Ripple Delete. Either way executes a "ripple delete" that deletes the portion of the clip you selected and moves *everything else* left.

Ripple, Roll, Slip, and Slide

Although you can essentially perform all of your edits simply by using the razor blade, the issue is what the rest of the clips are supposed to do in response to the edit you just made. In some cases, you don't want them to do anything. In other cases, you want them to move to accommodate the edit. You may want to drag in a short B-roll clip and slide it around to have it start and stop exactly where you want it to or change what the B-roll shows within those constraints. Most NLEs support Ripple, Roll, Slip, and Slide edit functions to handle these matters.

Resolve has several additional controls that improve on these basic edit functions. But many people get along just fine using the razor blade/Ripple Delete method—*Citizen Kane*, *The Wizard of Oz*, *Ben Hur*, and other Hollywood classics were edited with nothing more than a pair of scissors and scotch tape. If you do a lot of repetitive edits, Ripple, Roll, Slip, and Slide can speed things along, but it will take practice to learn how to use them efficiently. (These tools are covered in the Edit page chapter.) I guess I'm saying that not everybody needs the Ripple, Roll, Slip, and Slide tools, and getting good with these tools will take lots of practice. So don't spend the time to learn how they work unless they are necessary for your workflow. Occasionally I'll use Slip and Slide to finesse some B-roll into just the right place, but most of the time, the razor blade does most of the work.

Fade Up and Out

If you wanted to start with a fade up from black, you could use one of the effects generators to create a black slug (a 10-second clip that's nothing but black video), cut the slug to one second or so, and add a Cross Dissolve transition (Effects Library) to the first clip. That's the traditional method of getting a fade-up from black. But an empty track in Resolve is really an infinite black slug that's always there if you want to use it, and it's much faster.

Mouse over to the very top left edge of the first clip. A white marker will pop up. Mouse over the white marker, and double wedges will appear. Click and drag the double wedges. The marker will turn red as you drag it. The clip will now have a diagonal fade which is the same thing as a fade up from black. There's a handy timer that shows you the length of the fade. You can make it as fast or as slow as you want. You can do the same with the audio track.

Fixing Video Flubs

I have not seen your test footage (obviously), but I'm going to take a shot and guess that there's a segment in at least one of the clips that you'd rather not have there. Use the razor blade to cut both sides of the offending section and delete it (Ripple Delete).

Play the Timeline and look at the edit you just made. If it's a talking head, chances are, the clip will jump on this edit. This is the dreaded *jump cut*. There are many things you could do to reduce the clash, but one of the easiest is *Smooth Cut*. It's sort of a dissolve, but it also interpolates from the frames left and right of the cut to make the jump cut almost go away. Almost being the operative word. Smooth Cut works better with some clips than others.

Smooth Cut is in Effects Library > Tool Box > Video Transitions > Dissolve > *Smooth Cut*.

OpenFX effects can be previewed before you apply them to a clip. Put the play head on or near a splice. Go to Effects Library, select a transition (such as Smooth Cut) and scrub it back and forth. In the Timeline Viewer, you can see a preview of how that effect will look. If you like it, grab the icon and drop it on top of the offending edit. Play the Timeline again. (This transition may need to be rendered.)

The default timing of Smooth Cut is about one second. I find it works better if you shorten the duration (of course, it depends on the difference in the images of both clips), but IMO the faster you pull the band-aid off, the better. Zoom in on the Timeline and grab either the left or right side of the Smooth Cut icon, and drag to shorten the duration. If you just can't seem to smooth out the jump cut, try cutting off a few frames from one or both sides of the edit and apply Smooth Cut again. Sometimes, the glitches are caused by frames that just don't want to be there.

Instead of Smooth Cut, you could do a Cross Dissolve. You could also drop some B-roll onto video Track 2 to cover the jump cut (that's the traditional method). If you do use B-roll that way, it works better if there's no hint of an edit in the audio track.

Note that sometimes a dissolve or other transition may not "take." You grabbed a transition from the Effects Library, you dragged and dropped it onto the cut, and it's not there. If you are trying to do a one-second transition (such as a dissolve), there has to be at least one second of additional video embedded in the ends of both clips, even if you can't see it. Resolve has to have the additional video under the transition in order to perform the dissolve. If that extra footage simply doesn't exist, one way around that is to make a freeze-frame of the first or last frame of the clip, stick that on the end of the short clip and expand it to one second. The video under the dissolve is not moving, but dissolves actually work better when you are not dissolving motion to motion.

Finishing Up

Move down the Timeline from left to right, removing the parts you don't want, using only the razor blade for now. Use the Delete key to move everything left as you go. Tidies things up a bit. When you get to the end, you can add a fade to black.

Unlink the video from the audio when you want to work on either separately, and re-link them when you are done.

Save your work often. Or automate saves and backups by going to DaVinci Resolve > Preferences > User > Project Save and Load, and make sure *Live Save* is on.

Titles

Resolve has quite the sophisticated pallet for creating titles via Fusion. But for this exercise, we're going old school.

First, go to View > Safe Area and turn it on. This turns on various outlines that show you where it will be "safe" to put text. Almost no one watches TV on CRTs anymore, so the entire screen is technically "safe," but it's still a good idea to keep text in the traditional "Safe Title" area. Safe Title (inside box) meant that the text within those confines would be visible on 99% of home TVs (CRT type) despite the normal maladjustments. Safe Action (outside box) meant that whatever the actors were doing would be visible on virtually any TV if confined to that area.

From the Edit page, place the play head about where you'd like a title to be. Open Effects Library > Toolbox > Titles and select *Text* (the one *without* the plus sign). Drag the Text icon to the blank track right above Video 1. Resolve will create a new track (Video 2) with a blank five-second Text template. Click on the text clip and Ctrl+D to change the duration if you want (note the default presets for 1, 5, and 15 seconds).

You could have picked *Middle Lower Third* if that's all you needed, but Text lets you practice moving the title around. Select the Text clip and activate the Inspector (top right). Inside the Inspector, type your lower thirds text in the *Rich Text* box. You can pick the font, face, color, size—pretty much any text attribute you can think of. If you select the text clip, you can see a preview as you scan various fonts. Make it a lower thirds by changing Position Y to around 100 or so (be aware of where closed captioning appears).

Scroll down the list and take a gander at all the things you can do to fonts. Oh my. (Lucinda Grande was the broadcast go-to for years but is not available in Resolve for some reason.) For Drop Shadow, small numbers (Offset) work best. Same for Stroke (outline).

Broadcast and cable networks go to great lengths to identify the best typefaces to use with video, so copy their practices. In the distant SD past, we typically avoided serif fonts, particularly if the font was small. That's largely true even today (yet serifs would work fine in 4K). Common fonts for video include Helvetica, Source Sans Pro, Lato, Futura, Arial, and Gill Sans. For headlines, Futura, Gilroy Bold, or Bebas would also work well.

> **FYI**: You can fade titles in and out by grabbing the white marker at the top left or right of the text clip (which only appears when you mouse over the top corners). You can also click on the text in the Timeline Viewer and adjust its position or resize it. Or double click in the exact center of the on-screen text box and edit the text. Also, you can link the text to a clip which prevents misidentification later if you move things around. Note that the Safe Title line is a ledge where you can park text, and the location will be consistent from clip to clip. To keep text from getting lost, select the relevant clips, right-click, and *Link Clips*.

To keep your video from looking like a ransom note, once you've settled on the font, face, color, position, shadow, stroke, and other attributes, copy and paste the original text clip anywhere you need additional titles, and you only have to change the text in the box. Emerson may have said that "A foolish consistency is the hobgoblin of little minds," but he never said it to me.

Pre-fabbed Fusion titles are now accessible directly from the Edit page. If you want to get really fancy, there are animated title templates available for Resolve (Google them), and many of them are free. Most of the fancy titles you see on YouTube videos are pre-fabbed or store-bought.

Audio

You can work on audio using simple tools on the Edit page, or you can use Fairlight, a full-blown Digital Audio Workstation (DAW). My advice is to do one or the other but not both in the same project. For now, we're working with the audio tools on the Edit page.

Since at the start, we are fading up from black, you might want to fade in the audio as well. Grab the top left marker of the audio track, and drag it to create a fade like you did with the video.

Fixing Audio Flubs

There may be places in one of the clips where the audio has a tick or a pop—where the talent gestured and hit the mic, a lip smack, or a snort—in any case, something you want gone.

Since you listened to the clips, you probably know where these are. Often you can *see* where these audio anomalies are in the Audio Waveform in the track by visually scanning the audio track for places where the Audio Waveform peaks. Or you can just play the clip in real-time and leave markers by popping the M key every time you hear a bump or pop that needs addressing.

One way familiar to Premier Pro or Final Cut users is to set audio keyframes and "duck" the audio right at the point of the offending pop. To do that, expand the Timeline to get a more granular view of the audio track's waveform. Zoom all the way in (temporal slider). Find the bump or bang you want to tame. Hover the cursor over the volume control line until the cursor changes to wedges. Then press and hold the *Alt* key.

Now, every time you click the mouse, you'll set an audio keyframe. Move the playhead (mouse) a bit as you set each keyframe. You want to put one keyframe just before the pop, one after, and one in the middle. When you have generated three frames with the middle one centered on the pop, release the Alt key.

Graphically duck the audio by moving the center keyframe pip down a bit until the pop is not as jarring, but the narration isn't affected. But be careful about moving the keyframes on either end because that will change the entire track's volume fore and aft.

Setting keyframes is a reasonably effective way of reducing an occasional tick or pop (such as when the speaker at the podium bumps the mic or coughs or farts). Simply adding three or five keyframes, lowering the middle ones to create an inverted bell curve but not so much that the audio completely drops out, is usually a "good enough" fix.

A *better method* is to edit in a snippet of Room Tone (described in an earlier chapter) in place of the offending throat clearing or pop. That cleans it up completely. To replace an offending pop or cough with a piece of Room Tone, unlink the video and audio clips. Use the razor blade to remove the offending portion. Open the Room Tone clip in the Viewer, and set the I and O points so that the subclip will be shorter than the hole you just made (the I and O points can be just a frame apart). Grab the audio waveform at the bottom of the viewer (so that you are just pulling the audio). And drag the Room Tone subclip into the hole in the audio track. Adjust the length as necessary to fill the hole. Play the clip to confirm it works to your satisfaction. Finally, relink the video and audio clips so they don't become tangled later.

Adjusting Levels

Ultimately, you want the track to be as loud as possible *without ever exceeding the level where clipping/distortion occurs*. Audio that's too hot will splatter, clip, and otherwise distort on peaks, and that's usually not fixable (without ADR…). Simply dialing the volume down may work, but that almost always leaves a lot of dynamic range on the table. And taming a few peaks via the volume control results in a track with *low average volume,* which is not what you want.

There are tools in Resolve (particularly Fairlight) that will let you *raise* the average level of the track while preventing excursions/clipping and without audible artifacts. In the short sample video you shot for this exercise, there probably won't be very much variation in the volume—but with multiple mic setups or multiple speakers, there usually is.

Variations in levels and wide dynamics that are appropriate for the video (the speaker simply has lots of dynamic range) are fine. But it's more difficult to create a loud track without clipping when the audio levels vary wildly—you have to use more sophisticated tools than just the volume control. To be clear, the goal here is not to kill all the dynamics—those may be natural and appropriate. But I am saying that if the end game is a track with a high average volume and no clipping, the dynamics makes your job more difficult.

Earlier, you raised (or lowered) the volume of the tracks. If all of the audio tracks were recorded at the same time and place, you can probably just adjust the levels in the Mixer and be done with it. For most speakers/presenters, having the peaks hit between -15 and -10 dB is about right.

Whatever level you set, the volume must be adjusted so that no peak *ever exceeds 0 dB*. That said, it would be safer to give yourself a little headroom there and set a target of say -10 dB or -5 dB as the absolute max. In most cases, the goal for your final output is to have as much loudness as you can achieve without ever clipping or splattering.

Normalizing Audio Levels

Normalizing Audio Levels is one way to control volume. I'm not a big fan of *Normalizing Audio Levels* for every track/project because

I consider it a blunt instrument. Basically, you tell it what level you want the peaks to *never exceed*. The default is -9 dB. Normalizing Audio Levels will definitely stop any peaks from exceeding that (or any number you set), but it does so by lowering *the volume of the entire track* such that the level of the highest peak does not exceed the Target Level.

In many cases, the results may not be desirable. If you had just one well-above-average peak, the entire track's volume would be lowered simply to get rid of that one peak. In this case, you are using a one-off anomaly to set the volume of the entire track.

That can be useful sometimes, particularly with music, and it certainly is quick. But it's using the *dynamic range* to control *volume*. If you control peaks simply by lowering the volume, you are potentially left with a weak audio track.

If you want to try it, you can activate *Normalizing Audio Levels* by right-clicking on any audio track. The number you are setting is the Do Not Exceed value. The Target Level for a talking head presenter should be such that the peaks only occasionally hit -10 dB and never get to 0 dB. The default value does that. Normalizing Audio Levels could be done as you import clips instead of adjusting individual tracks using the In-Track Volume Control.

Audio Effects

There are many audio effects tools available from the Edit page without having to use Fairlight. For most social media-type projects, the Timeline tools are probably enough and are easier to use.

You can get to a four-band EQ by activating the Inspector > *Audio* and scrolling down till you find *Clip Equalizer*. The Inspector defaults EQ to off. Turn it on.

For just about any audio shot on location, you could use *Clip Equalizer* to roll off some of the low end (~80 Hz) to reduce rumble. Or you could lop off the extreme high end to reduce any hiss or excess sibilance. In any case, you only get four bands with this EQ. You can also pan from the Inspector if you want (if the tracks are stereo).

Later, we're going to cover some of the Fairlight audio effects, which are now available on the Edit page under Effects Library > Audio FX > Fairlight FX.

Go ahead and cut the clips, adding whatever Video Transitions you desire. Experiment. When you are done, play the completed Timeline from end to end and look for anomalies.

Picture Lock

Before exporting your project, make sure that your first clip is flush with the start of the Timeline (Home and Ripple Delete if necessary) because an empty space in the Timeline is interpreted as black video. And zoom out of the Timeline and make sure there are no loose pieces left lying around.

At this point, you have done everything to the video and audio tracks you are going to do. To prevent any possible slip-ups, make sure all video and audio tracks are locked using the Lock icon (on the Video and Audio tracks), so they can't be accidentally changed later.

Deliver

At this point, the deed is done. Time to .mov on.

Mac vs Windows

While there are few differences in Resolve between the Mac and Windows versions (except for obvious things like Command vs. Control, etc.), there is one very important difference between Macs and Windows when it comes to Deliver—ProRes. ProRes is an Apple product, and it is not licensed for the Windows version of Resolve—so exporting in ProRes is not an option in Windows. (You can import and edit ProRes on a Windows machine, however.)

Deliver Settings

Go to the Deliver page and choose either YouTube or Vimeo. Unless you log into your account, the video won't be posted (there's also a check box to inhibit posting). The default values are pretty standard. You can change them if you want. (There are lots more details about compression and encoding in the chapter on Deliver.)

For any of these paths, you'll need to give the output file a name and specify a location where *your copy* will be stored on your computer. You can select a Format, Codec, or other attributes, but the defaults are probably OK for now. One exception would be the encoder (now rather obscurely called "Type" for some reason). If your GPU has a hardware encoder, set your graphics card as the encoder "Type" rather than Native.

In Deliver, *Add to Render Queue* simply adds the project to the queue. You have to *Render All* to get the show going. There are two indicators on the Render Queue panel that provide a guesstimate of how long all this is going to take, and they never seem to agree exactly. You can also watch the play head move through the Timeline as the final render progresses.

When you are done, locate and play the finished file (local copy) to make sure it's exactly like it's supposed to be.

From the Cut, Edit, Fusion, or Color pages, you could export a project to a file using File > *Quick Export*, but you won't have a lot of say over how that file gets created. You can even publish to YouTube and Vimeo using this method, but really, it's typically used for making quick H.264 files to put on a local server or SD card for review purposes.

Last Word

Even though you completed this exercise, I doubt you would claim to have mastered it. The first time I used Resolve to do a similar project, it took me more than two hours (and I've been editing for years). The second time took about an hour, and the third, fourth, and fifth times, much less.

I strongly encourage you to shoot some more video and repeat this exercise *several times* until you really, really get it. Then you can feel confident enough to move on to the next exercise.

The primary reason for this approach is because some of the controls for a single function are in two or three places and have odd names. If you only do it once, you will not remember how to navigate the secret passages. Resolve can seem frustrating at times because of the many different ways you can do any one thing and the many different (and obscure) places Resolve hides its charms.

When you finish an exercise, *always review the output video file* to make sure everything's working as it should, then start again from scratch (New Project, Reset UI Layout, Media page…).

When you have finished, I would also advise you to return to the Edit page, go to Playback > *Delete Render Cache*. Renders create huge files, and if you don't delete them when you complete a project, they will quickly fill up your cache drive. Also, delete any Optimized Media or Proxy Media files if you used them.

Exercise Due

For this next exercise, you're going to tackle Chroma Key (AKA *green screen*) by learning how to use the tools on the Color page and Resolve's (in)famous *Nodes*. For this exercise, you should shoot another talking head, this time seated in front of a green screen.

Don't have a green screen? You should. A green screen, even a cheap one, lets you do some amazing things with Resolve. I bought a green screen kit with two soft lights from Amazon for under $100. It's not that sturdy, but it works. If you already have two soft lights, you can get a cheap green screen with a stand for as little as $11 on Amazon.

If you've got a spare wall, you could paint it. Genuine Chroma Key paint from film and TV supply houses such as Rosco costs about $50 *per quart*. But it's the real deal made from single-pigment paint and is flatter than hardware store flat interior latex. Behr's "Sparkling Apple" or "Green Acres" from Home Depot are reasonable substitutes. Sherwin-Williams' "Neon Green" is also very good. If you do use hardware store paint, make sure it's *flat*, and put it on a smooth surface using a roller. You could even get a sheet of Styrofoam and paint it with hardware store paint for a workable, portable green screen.

If setting up a green screen is simply out of the question, you should be able to find some green screen footage online for free (as long as you don't plan to distribute it). Doesn't matter if it's watermarked. You may have to do some searching to find a downloadable green screen clip with a talking head or person standing in front of one, but they're out there, and you only need a few seconds. In fact, you could even use stills for this because you are really only working with one frame anyway.

You'll also need a background image such as an office interior—something to key over. Choose a background image that's sharp and not already blurred. And it should be the same resolution as your camera footage (or a larger). I found some really nice free backgrounds on youtubestock.com, but a Google image search was also productive. Don't worry about royalties or watermarks unless you are planning to distribute it.

Shooting a Green Screen

There are several points to consider when you shoot a green screen. First, the green screen usually doesn't need to fill the entire frame. Most of the time, you only need enough green screen to cover the talent and their expected range of motion. For seated talent, that's not very much. The areas outside the green screen are going to be masked off with a "Garbage Matte" anyway.

The green screen should be evenly lit with soft lights such that neither the lights *nor reflections from the green screen itself* spill onto the talent. Likewise, the talent lighting should not impinge on the green screen either.

You can tell if the green screen is properly lit by shooting a sample, bringing the clip into Resolve, and looking at it on the Waveform scope. The green line on the Waveform scope should be flat. It should also be thin. The flatter the line, the more evenly lit the scene. The thinner the green line, the more pure the color. Play around with the lights until you achieve an evenly lit screen, as shown on the Waveform scope. The green screen should not be lit too bright. A Luma between, say, 384 and 640 is probably right. If you are using a fabric screen, iron it (or better yet, steam it) because wrinkles don't help.

The talent should be as far from the green screen as possible. In practice, that may not be more than 6-10 feet, but you need some physical separation because you don't want light from the green screen to spill onto the talent (or any foreground object) because it will leave a green cast. Much of the need to "clean up" a green screen shot is getting rid of this spillage. The less spillage you have in-camera, the less clean-up you'll have to do. And obviously, the talent should not be wearing anything remotely green (including makeup).

And it will make things much easier if you eliminate any fiddly bits in the foreground that will be in front of the green screen. Solid shapes evenly lit require less touchup than intricate objects with delicate tracery. These are concrete examples of shooting it right so that you have less work to do on the back end.

Background Plate

If you are shooting 1080p, try finding a large image for the background (1080 by 1920 pixels or higher). Obviously, if you're shooting 4K, the background image would need to be even higher resolution. You may need still higher resolution for your background plate because the format may not be 16:9 (or whatever your video is) or because you may want to zoom into the background plate or move around inside it for esthetic reasons.

Since the foreground frame is horizontal, the background image should be horizontal as well. You could use video for the background, but for this exercise, stills are easier to practice with.

I found a beautiful background shot, but, unfortunately, it had a model right where I wanted my talent to be. Since the image size was quite large, I was able to use it by moving around in it (via the Inspector) and using the foreground talent to mask out the model in the background.

Convincing Composite

Although sometimes we use green screen simply to provide a background with a pattern or some other image besides our garage or basement walls, the principal reason is to convince the viewer that the talent is actually somewhere else.

The real trick is picking an appropriate background and tweaking both the foreground and background video tracks so that it looks real. That's the skill level you want to try and achieve, so for this exercise, pick a background plate suitable for the foreground image and light the foreground appropriately.

If the background is a dark alley, you'll have to shoot the foreground video such that it will look realistic when composited. If there's a strong warm key light coming from top right in the background, then light the foreground to match. And by "match," I don't mean an exact match but lit such that the resulting composite image is plausible. Resolve has tools on the Color page that will let you correct both video tracks and composite them seamlessly, but it's better to give yourself a lead by shooting the foreground in such a way that it's believable that your talent could actually be in front of the background image you've chosen.

Green Screen Exercise

Create a New Project and import your talking head green screen clip and the background plate still to the Media Pool, then go to the Edit page.

But first, a few (new) words…

In creating a green screen, you'll also be creating an *alpha channel*. Normal video can be thought of as three channels—R, G, and B that vary in Luminance. The alpha channel is just one more, but it's B&W and binary. The job of the alpha channel is to hold the *shape* of the key. It does not carry any Luma or Chroma information. The alpha channel holds the *shape* of everything green which allows Resolve to swap that portion of the alpha channel with some other image. This process is called Chroma Key or, more often, just "green screen."

Speaking of Chroma Key, *key* is an old TV term that means replacing one part of an image with another. Chroma Key uses the *Chroma* signal (color) to tell the keyer (which used to be a rack-mounted device) which areas to key (replace) and which to leave alone. The BBC still calls it a "CSO" (colour separation overlay), AFAIK.

In the US, it's often called "green screen" because green is the color most widely used in video. Technically speaking, blue is the opposite of skin tone, but there are more blue suits than green ones, and more "suits" appear on TV than in the movies. That said, Hollywood typically used blue screen when working with film because the color they chose for the blue screen was exactly the opposite of skin tone and easier to work with using just chemical dyes and color filters. TV historically preferred green over blue because the green video channel was used to generate the black and white alpha channel. That was because the green channel had much more detail and less noise than the blue channel (not true today).

What you are going to do in this exercise is create a Chroma Key/green screen shot by creating an alpha channel image whose contents will be defined by a single color in the foreground video (the green screen itself). The solid black areas of the alpha channel will be keyed out by the keyer (software in this case) and replaced by another video (background video) to create the resulting composite image.

Background Plate

In this exercise, when compositing a green screen (creating one clip from many), *the background image must be on Video 1* and the foreground (green screen) footage on Video 2. Basically, the green portions of Video 2 become transparent to the image on Video 1.

Shoot a green screen clip. For learning how it works, short clips are best…

Drag the background plate (still or video) from the Source Viewer to the Video 1 track and only grab the video from a clip if you are not using a still (by grabbing the film-frame icon near the bottom of the Source Viewer). Ripple Delete so the background image starts at 00:00 on the Timeline (you could slide it there as well).

If the background plate is not full-screen already, open the Inspector, and using *Transform*, zoom in on the background plate (if necessary) so that the background image is full frame. If you have enough pixels in the background plate (higher resolution than your Timeline), you can also move around within the image so that it will fit (visually) behind the talent (so that objects in the background are not growing out of the talent's head, for example).

EXERCISES 125

Foreground Plate

Drag the foreground (green screen) video to the area above Video 1 (thus creating Video 2).

Creating Video 2 means there's an extra audio track (Audio 1) you won't be needing. Delete it by right-clicking on Audio 1 and selecting *Delete Empty Tracks*. Resolve is named for DaVinci, not Dali, so there's only so much you can do to create space to work in. It's mandatory with Resolve to delete tracks you don't need and kill panels you aren't using.

Pulling a Green Screen

Now switch to the Color page. Select the thumbnail of Video 2 (green screen shot). A default empty Node will already be there. I assume you've already shot a black and white reference as explained in Exercise *Uno,* so go ahead and create a WB/LUMA grade for the foreground clip as you did in Exercise *Uno*. Then, find a "hero" frame that's representative of the entire clip.

Repeat for Video 1, but if you didn't shoot it, you won't have a reference for white and black. However, the scopes will allow you to tame any excursions.

In the Timeline (Edit page), grab the right side of the background image and drag it until it's the same length as the talking head clip. Set Timeline Proxy Resolution to Half resolution and Render Cache to Smart.

Select the green screen clip. Go to Color > Nodes > *Add Serial Node* (Alt+S). Right-click on the Node and label it "Chroma Key." In the dark gray grid where the Nodes dwell, right-click anywhere and select *Add Alpha Output*. A blue dot appears. Notice that there's a blue dot on the Chroma Key Node as well. Connect the dots. (Just the one from the Chroma Key Node to the alpha channel.) You've just created an alpha channel.

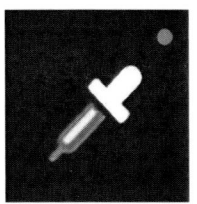
Qualifier

Click on the Qualifier eyedropper to activate (center of screen). The default setting is HSL (Hue, Saturation, Luminance). You can work in that mode if you want, using the eyedropper to pick individual green pixels to tell Resolve what color you are trying to key out. But I think you'll find it easier to switch to 3D (eyedropper icon option in the Qualifier).

Invert Key

The 3D mode lets you draw *lines* to select the green pixels—hundreds and hundreds of them at a stroke. Under Qualifier 3D, select the eyedropper plus (+), so you can draw more than one line. You can start and stop, too (lines do not have to be contiguous). You'll find that the more green pixels you identify, the easier it is to pull a clean green screen. So draw all over every nook and cranny of the green screen (particularly the edges of the talent or objects). Just make sure you don't color outside the lines. Ctrl+Z if you do. You can zoom the Viewer in and out using the mouse wheel if you need to see details better. Because of the improvements in Resolve 18, most PCs will be able to show you in real-time what your selection is doing (and you'll be able to see which areas need attention).

When you release the eyedropper, the image in the Timeline Viewer will be keyed. Although, it will be the reverse of what you want. Next to the eyedropper plus icon is the *Invert* button. Click it.

Check the results using the full-screen monitor. The resulting green screen composite will almost certainly have areas that need further attention. If there are green screen areas that didn't get picked up, use the 3D Qualifier again and carefully pick them out. Once the green screen looks good, there are additional tools you can use to clean it up further.

Highlight B/W

It may be hard to tell which areas are keying badly and why. At the top left of the Viewer is the "Magic Wand" Highlight icon. Activate it. To the far right of that is a Highlight B/W icon. Activate that one too. (B&W is now the default in Resolve 18.)

Highlight

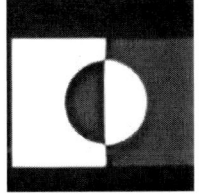
Highlight B&W

Now you can see the image as a black and white *alpha channel* representation. Go to Selection Range > Matt Finesse and adjust *Clean Black* and *Clean White* until you have (virtually) no white pixels in the black areas and (virtually) no black pixels in the white areas, or you have at least reduced them—trying to eliminate every last one with these controls may cause unwanted artifacts. In other words, don't get jiggy with it.

Depending on the image, values above 10 may start giving your talent a strange edge glow, so watch carefully and adjust deftly. If more extreme values are needed, those should have been dealt with using the 3D Qualifier or lighting. But you don't have to remove every errant pixel either.

Turn Highlight off and zoom in. Add some *Denoise*—around five seems to be the convention. (Resolve 18 moved this control to Matte Finesse.) *Blur Radius* softens the edges while *In/Out Ratio* (usually negative numbers work well here) cleans the edges up (known as a *choke* in printing).

Try turning Blur Radius up a bit (start around 3 or 5 and see how it looks—don't blur so much that the details drop out). Turn the In/Out Ratio up and down. When adjusting Blur Radius and In/Out Ratio, watch the edges carefully on a full-screen monitor. (You could even turn Timeline Proxy Resolution off at this point for greater resolution.) It is useful to move a control's value through its full range while watching the monitor *to see what the control is actually doing.*

The goal of these adjustments is to minimize the defects around the edges. The proper setting is always a compromise between taking out a defect vs. adding an artifact. Pay close attention to the talent's hair. When you are done adjusting, you should have a rather clean green screen composite. You may notice that the control scheme is similar to After Effects.

Go back to the Qualifier and slide *Despill* while watching the monitor. Voila!

I would have told you about Despill first, but it's such a powerful and effective tool that some people neglect setting the Matte Finesse controls because Despill works so well.

Garbage Matte

Since the green screen you shot does not cover the entire background (if you shot it as instructed), you'll also need to create a Garbage Matte. It's called a "Garbage Matte" because it's a "down and dirty" way of removing large chunks of an unwanted image without too much effort.

With the Video 2 (foreground) clip still selected, select the Chroma Key Node, then go to Color > Nodes > *Add Serial Node* (Alt+S). Label this one "Garbage Matte." The Garbage Matte Node should be downstream of the Chroma Key Node.

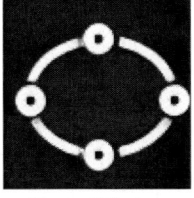
Power Window

Delete the existing alpha channel connection (highlight the dotted line and Delete). Now connect the alpha channel of the Chroma Key Node to the alpha channel input of the Garbage Matte Node and then to the alpha channel output (blue connectors).

Curve Pen Nib

Select the Garbage Matte Node. On the control pallet in the middle of the Color page, you'll find the *Window* icon. Click on it. Select the Curve pen.

To define what the Garbage Matte is going to mask out, you are going to draw a box where the green screen would be if it covered the entire background. Zoom way out so you can see the off-screen border. Go ahead and draw a single box in an appropriate area (by clicking to set the corners of the box). Doesn't have to be perfect—you can adjust it later.

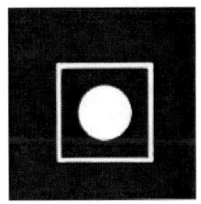
Invert

Note that in Window, to the right of the pen highlighted in red, there are two control buttons. These used to be labeled "Invert" and "Mask" in previous versions of Resolve, but, for some reason, the names got dropped in Resolve 16 and didn't get picked up in 17 or 18. Click on the Invert icon. The Garbage Matte should now be blocking the non-green screen area you described with the box.

Mask

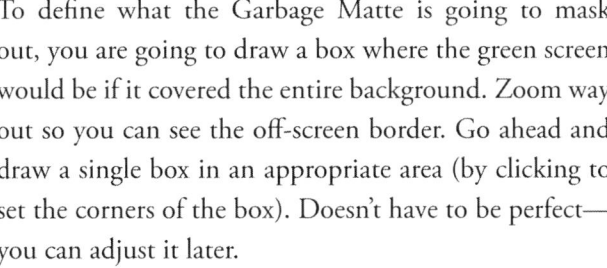

EXERCISES 129

You can draw additional boxes if you need to by clicking on the *plus Curve* pen tool (center).

If you use more than one pen, click *Invert* for the first pen and *Mask* for any additional ones. (Don't ask.)

Once drawn, you can grab the box's wireframe and move it around if you need to. As long as the box covers the area you want, it doesn't have to be pretty—that's why it's called a "Garbage Matte." Also, you could have chosen a box (Linear), Circle, or another shape to define the matte if that would have worked for you

Adjusting the Background Plate

Back on the Edit page, select the Video 1 clip (the background plate). If the background plate is a real-world image, it would normally be a tad out of focus (as if your camera has limited depth of field). To fake that, go to Effects Library > OpenFX > Resolve FX Blur > *Gaussian Blur*. Drag the Gaussian Blur icon onto the background clip. An FX icon will appear on that clip.

Open the Inspector and select Effects > Open FX. Reduce *Horizontal Strength* until it "looks right." (Vertical and Horizontal Strength are ganged together at this point.) Not only does a little background blur give the composite sort of a "film look" because of the depth of field, a blurred background makes the foreground look sharper. The default is .4, and you'll probably want less than that for realism.

You also could have done this from the Color page by creating a Blur Node for Video 1 clip and adding the Blur effect.

3D Keyer

The above method is the tried and true method of Chroma Keying. But now there's an easier way to Chroma Key. It's called 3D keyer, and it's available as a Resolve FX "Filter" from the Edit page. Start over from scratch.

On the Edit page, go to the Transform control at the bottom left of the Viewer and select *Open FX Overlay*. This is important. *Do not skip this step.*

Select Video 2 (green screen shot). Now go to Effects Library > Open FX > Filters > Resolve FX > Resolve FX Key and drag *3D Keyer* to the Video 2 clip.

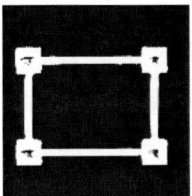
Transform

Open the Inspector and select *Effects*. You'll get the same controls you just learned how to use, but the whole process is on one page—the Edit page. The eyedropper will already be active (but in various builds, it displays the arrow cursor instead). Use the cursor/eyedropper to draw a line on the green screen. Immediately (or eventually, depending on your graphics card), the screen will key. Click on the eyedropper plus (+) if you need to add additional lines. You should be able to tell where you need to do that. Draw more lines in areas that are not keying cleanly.

In the Inspector, scroll down to Output and select *Alpha Highlight B&W* to adjust the Matte Finesse controls. You should be able to easily see what the controls are doing.

When you're finished, change the Output to *Final Composite*. Under *Keyer Options*, select Despill and unselect *Show Paths*. Play the clip and tweak the Matte Finesse controls as you see fit.

Creating a Garbage Matte via the 3D keyer on the Edit page requires a slightly different technique. Go ahead and try creating a Garbage Matte in the Inspector. You only get the choice of a rectangle or an ellipse, but most of the time, that's all you need.

I should mention there is yet another way to Chroma Key via Fusion, and it's covered in the Fusion chapter.

Grade Your Own Work

Now it's time to try your hand at grading. This exercise is not going to make you a colorist, but it will give you some feel for what you can do with the Color page in Resolve.

On the Color page, select the thumbnail of the background clip. Make sure the WB/LUMA node is selected, then Color > Nodes > *Add Serial Node*. Label this Node "Grade."

Bring up the scopes (Ctrl+shift+W) and select Waveform. What you are trying to do is identify the brightest parts of Video 1 and raise or lower Gain on Video 1 to match/compliment the foreground image.

Viewers usually expect the foreground to be lit hotter than the background (it usually is), so the background will naturally have more shadows and a lower Gain and Gamma too. But it really depends on the subject matter of the foreground and background images. In all cases, use your eye to judge the work.

Next, look at the color cast of both images. The background plate I chose had a strong yellow cast. I dialed that back a bit by moving the Gamma color wheel a bit toward blue. I remind you again that the circled arrow on just about every control resets that control back to the default (as does Ctrl+Z) in case you need to get back.

Typically the Gamma Color Wheel should be adjusted on both video tracks to bring the colors more in line with each other, but that, of course, depends on the images—bright colors may require adjusting the Gain Color Wheels as well. Saturation may need to be adjusted on both clips as well.

Create a Grade Node for the Track 2 foreground clip just like you did for the background and do the same thing—except this time, create the new Node *after the Garbage Matte Node* (if used) so as not to affect the green screen (the alpha channel does not need to be disconnected/reconnected for this step).

Bypass Effects

You (usually) don't want to grade the foreground video *before* the Chroma Key Node because you are using the Chroma values to key on. But once they've done their job, you can change the Chroma values (and any other values) to anything you want. Adjust the Gain, Lift, and Gamma color wheels on the foreground clip to realistically match/compliment the background. The Waveform scope will help you see what you are doing here, and the full-screen Viewer will show you if you've done it. You may need to go back and forth to get it *just so*.

Depending on your computer's CPU/GPU, it could take several minutes to render these effects. For that reason, use short clips for these exercises. I also want to point out that there is a handy button at the top center (a color icon with stars—and I'm guessing that's the Southern Cross since the BMD folks are Aussies). That button lets you bypass all the Node grades and effects you applied on the Color page. You can also turn any Node off (or back on) by clicking on its label (another reason for labeling all Nodes). It doesn't turn off Resolve FX effects applied on the Edit page, however.

Compound Clip

You currently have two video tracks (green screen and background plate). If you wanted to fade up from (or down to) black, it's hard to get both tracks to fade in and out at exactly the same rate. One solution is to combine (composite) the two video tracks into one and then fade the composite.

From the Edit page, shift-click on both video clips, then right-click and Link Clips. Select New *Compound Clip* which will merge the two clips into one composite. Name it something and click *Create*. Now the two clips are one, and you can grab the white marker at the beginning and fade the single composite clip in and out. The foreground and background clips will now fade at exactly the same rate.

It should be pointed out that once a compound clip has been made of two videos, the compound will then be treated as a single video clip (which it is). This means you can add effects to the compound clip, and the entire composite image will be affected. You should color correct and grade the foreground and background footage separately, but you can still grade the composite clip later, as a whole, if you want. You can un-composite a clip by right-clicking on it and choosing *Decompose in Place*.

Go ahead and fix the audio as in Exercise Uno and output the project using Deliver. Play the output file and make sure everything you just did works.

Practice, Practice, Practice

At this point, you've used a lot of new techniques, but I doubt you would say you're an expert with only one rep under your belt. As with the first exercise, take the same clips and repeat Exercise *Due* several times. Repetition is the only way to master it.

Exercise Tre

If you've repeated the first two exercises several times, then at this point, you should have a fairly good grasp of the Resolve UI and its control scheme. So, let's step it up...

Color Grading

Color Grading is, without a doubt, the Big Kahuna. In fact, it's Resolve's main claim to fame, and to this day, no NLE does it better. The original da Vinci system was a very expensive hunk of hardware married to some rather clever software that graded (color corrected) digitized, faded, scratched prints of old films. Over the years, as PCs and Macs got a lot more powerful, it wasn't necessary to use a minicomputer anymore (ask your grandparents), and da Vinci was ported to a software-only system that could run on PCs and Macs. At this stage of the game, DaVinci Resolve, to give it its full name, is the go-to for feature films, broadcast, cable, and streaming. And now, you can take advantage of the power and elegance of Resolve's grading, too.

Grading means many things to many people and ranges from something as simple as basic color and exposure correction to punching up flat video, matching clips shot in different lighting, matching B-roll to the master shot, all the way to completely relighting the scene.

One of the main uses of grading is to match takes shot on different cameras under different lighting conditions, often on different days, so that the differences (particularly when checkerboard edited) are not so pronounced or even noticeable. In fact, the WB/LUMA Node you create as the first Node begins the grading process by setting the color to "normal" and white and black levels appropriately.

Since grading often means working with isolated areas within a frame, being able to isolate objects and apply effects to specific parts of a frame is a key skill. In the exercises that follow, you'll learn various techniques for isolating objects/areas and applying typical effects—not just grading.

> **ASIDE:** Back in the day, when all of this kind of work was done on filmstock using chemicals, producer Gene Roddenberry was shooting a pilot for a sci-fi. He had the actress who had been hired to play a "moon maiden" made up with a slightly green cast (she was supposed to be an alien, and the cliché—then as now—was that aliens were various shades of green). The crew shot some test footage and sent it off for processing. When they got it back from the lab, the actress's face showed nary a trace of green. So they used much greener makeup and reshot. The film came back, and this time there was a very slight green cast, but not as much as they'd hoped for. For the third and final time, they used an intensely green paint. The next day, the lab called and told Roddenberry he needed to get a new DP—the lab was at the limit of what it could color-correct with the footage he sent them. The sci-fi, of course, was *Star Trek: The Original Series* (TOS).

Now down to business. But before we go any further, I do need to drop a stipulation: I don't know how much of this you can learn from a book. Color grading is based on what the frame *looks like to you*. Grading means creating a visual look that fits your project, and only you know what that vision is. It could be something appropriate for social media, or it may be a particular film look you are going for or some highly stylized effect. In this exercise, I'm just going to show you how a grade might be done—the specifics of how you intend to use grading is up to you. This exercise won't make you into a professional colorist, but it will give you some basic experience with the tools. From there, you can go wherever your eyes take you.

Grading the Clip

For the first exercise, shoot a clip with a face in it (head shot). You can probably reuse the footage from Exercise *Due*.

Create a New Project and import media. White balance and set Gain and Lift as in Exercise *Uno* (ending up with a WB/LUMA Node). If you are using H.264/265 footage, it's probably a good idea to generate Optimized Media or Proxy Media for this exercise (and work with very short clips). Needless to say, when grading, you may want Timeline Proxy Resolution off so you can see more detail and also with the Viewer zoomed in (or full-screen).

On the Color page, turn on scopes. For what follows, you'll need the Waveform monitor and the Vectorscope. We haven't covered the Vectorscope yet. It got its start during the old NTSC days (broadcast TV) to standardize color. The Waveform monitor displays brightness (Luma) while the Vectorscope shows Hue and Saturation (Chroma).

You can think of a color *vector* as an angular position somewhere on a circle. *Where* a line or dot (a smear is more like it) appears on the Vectorscope (within a 360-degree circle) determines its Hue. In fact, the Vectorscope is marked with boxes for R, G, B, plus Magenta, Cyan, and Yellow. How far the line or blob extends from the center determines its Saturation. The combination of Hue and Saturation defines Chroma. (If you want to truly understand how a Vectorscope works, shoot a box of watercolors or a collection of brightly colored children's toys.)

Back in the day, we would record color bars at the beginning of every videotape. Prior to editing or putting the tape on air, we'd play the videotape and watch the Waveform monitor and Vectorscope. We'd set white and black levels, then adjust the Chroma phase until the color bars were in the correct boxes on the Vectorscope. That way, we could be reasonably sure the colors at home would match. (Even so, NTSC was often said to mean "Never the Same Color.")

(If you want to see how this works, go to Effects > Generators > Color Bars and pull the SMPTE color bar clip into the Timeline and look at it on the Vectorscope.)

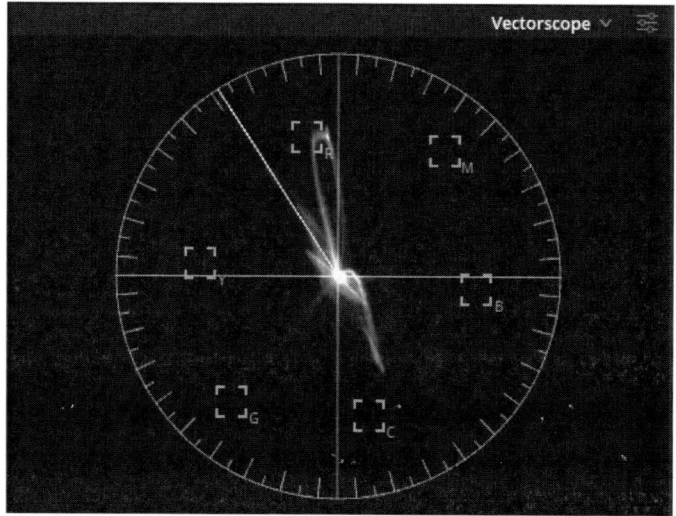

The Vector scope represents colors by their "vector" or angle

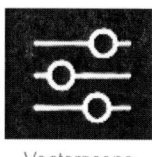

Vectorscope Sliders

Open the Vectorscope controls (three sliders) and turn on *Show Skin Tone Indicator*. A dotted line will appear indicating where skin tones should be. Notice there are not different lines for different colors of skin. We are all the same color. Any differences are simply a matter of Saturation and Luma, not color.

Mixed Lighting

Historically, when all we had was incandescent (tungsten) lights, interior lighting was warm (reddish), while exterior lighting (sunlight) was cool (bluish). This is another instance where we use words in reverse of their true meaning. Red light is actually cooler than blue light temperature-wise but for some reason we call red light warm and blue light cool.

Technically, the color of light is based on its temperature measured, for historical reasons, in degrees Kelvin. Tungsten light is ~3200 Kelvin while daylight is ~5500 Kelvin. (Thus, blue light is actually hotter, not cooler, than red light.)

I'm sure you've seen differences in tungsten and daylight sources. In some cheaper feature films, when a character opened a door, the exterior was lit with startling blue-white light—sort of like *Close Encounters* but without Spielberg. We used to have to put large sheets of orange gel (Wratten 85B) over the windows to warm up daylight in order to match the warm interior tungsten lights. The obvious solution would have been to shoot the interiors with "daylight" lights, but tungsten lamps don't make much blue light, so at the time, it was easier to filter the blue out of sunlight and shoot for "tungsten" than the other way round.

With CFLs and LEDs, you can now easily shoot interiors with "daylight." But it's still possible to end up with footage that has warm, low contrast interior lighting and cool, high contrast exterior lighting in the same take. In some cases, that can be acceptable or even desirable. In non-dramatic, non-stylized videos, it usually isn't.

For this grading exercise, load your project footage and get back on the Color page. Also, create two new serial Nodes and label them Grade and Face.

> **CAVEAT**: If you need to add a lot of *different* corrections and effects to a clip, create separate Nodes. Putting unrelated corrections or effects into a single Node can mean having to completely redo the Node in order to make slight changes. I usually end up with between three or four Nodes when grading a clip (but once done, the Nodes can be copied from that clip to similar clips).

For this exercise, you'll need the WB/LUMA Node to normalize the exposure and make sure your highlights and shadows are within acceptable limits. Then you'll need a Grade Node to make any macro color corrections to the full frame. This is nothing you could not have done in the WB/LUMA Node, but separating these corrections is usually a good idea. That way, you can apply the same grade to shots that need to have a different WB or exposure correction (or vice versa).

Select the Grade Node and adjust the color of the Offset and/or Gain wheels (and possibly the Gamma and Lift color wheels) until the color cast of the overall frame is to your liking (it may already be). Since I shoot Sony and not Canon, I almost always have to boost the Saturation a bit. A more subtle way is to raise *Col Boost*. Col Boost increases saturation in just the low sat areas, which lets you boost saturation just in the areas that really need it without making the frame look cartoonish. I find it helps to have a full-screen monitor up for this.

> **IMPORTANT** When grading a green screen clip, wait and grade the clip *after the Chroma Key Node* so as to not affect the key.

Skin

Switch to the Face Node. To grade/fix just the face, the first thing you'll have to do is isolate the face (otherwise, it's just another full-frame Grade Node).

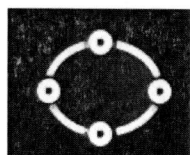
Power Window

Zoom in on the face. Select the Power Window controller and the circle icon. Adjust the location, size, and shape of the Window until the *thick* white line encompasses the entire face. You do this by moving the *solid* white dots. The thin white lines are the "feather," meaning any effect you apply won't stop abruptly at the edge of the thick white line but will be feathered or blended. You can adjust the range of the feather by moving the red dots. Avoid catching any hair, background, or clothing inside the area described by the thick white line (do what you can).

With the face defined by the Window, any effect added to the Face Node will only be applied to the area inside the thick white lined circle (which you probably made into an oval) with just a touch in the feathered area. To see this, dramatically shift the Offset Color Wheel (then reset).

Face with Window

There are as many techniques for fixing faces as there are YouTube channels with Resolve tutorials. One I particularly like is by Sidney Baker-Green. Baker-Green has a simple "formula" that works for all skin tones and is easy to duplicate. (These corrections should be subtle so, on the Primaries panel, change from Wheels to *Log*–which gives you much finer control.)

Baker-Green always shifts the Midtones a bit towards orange and the Shadows a tad towards red. There are technical reasons why this almost always works. (Baker-Green has an online color grading course that's just the ticket if you really want to get good at this. Check out shopsidneybg.com for details.)

Also, check out Waqas Qazi's interesting video where he explains how to get the "Orange/Teal" look used on many movies and network TV programs.

Check the Skin Tone Indicator on the Vectorscope. Carefully adjust the Offset color wheel until the skin color of the talent lines up more or less with the Indicator. In some cases, you might not like an exact match. It's OK to cheat. Ultimately, it's what you see that matters. I often have skin tones that are slightly redder than the Indicator suggests, which I think is fine, so I leave it alone.

Next, move the Midtone wheel towards orange and the Shadow wheel towards red. When you achieve a skin tone you like, adjust the Saturation until it looks natural. Now you should have nice skin tones in the face.

The problem with this method is that faces sometimes move. To get your Window to move with the talent's face, you'll need to use Tracker. Fortunately, it's only one button over from Window.

Select Tracker.

Tracker

Tracker basically acts like a *camera*. You can tell Tracker to Pan and Tilt to follow the talent. You can let it Zoom and Rotate if necessary, and if all else fails, you can let it track in 3D space.

The problem is that each degree of freedom (Pan, Tilt, Zoom, Rotate, and 3D) requires separate calculations. And unless they're skydiving, the talent is probably not moving in all those dimensions at once, so calculating in five dimensions is usually unnecessary (and shouldn't be the default, IMO). Having Resolve try and follow the talent in five dimensions makes it more likely Tracker is going to lose track (fail to follow what you've designated) in addition to taking a very long time to render.

Tracker Control Ribbon

To speed things up, turn 3D off. If you could (in your mind's eye) follow the talent just by panning and tilting, only leave Pan and Tilt on. For seated talent (limited range of motion), turn off Rotate and Zoom as well. (Although, if they move to and fro, you will need Zoom.) Add additional dimensions only if you need them to get Tracker to track. 3D seems to cause the most problems—but 3D is sometimes necessary to get Tracker to track properly.

Tracker Keyframe

Still on the Face Node, set the play head to the beginning of the clip. In the Tracker control panel, press Track Forward on the *Tracker's transport controls*. With only two or three dimensions to deal with, Tracker will probably run at the correct FPS (or nearly so) on most machines. As Tracker does its thing, you can watch it move the Window around to follow the face.

When Tracker has finished, the Face Node's Window should have followed the talent's face as they moved. Play the clip (using the Viewer Transport controls) and make sure it works. You could also scrub the Timeline to make sure it's tracking accurately. (Note that the "play" button on Tracker is a *Track Forward* control, not Play.) Now, any corrections or effects you add to this Node will only apply to the area delineated by the Window (which, if you did everything right, will just be the face, despite it moving around). You can also adjust the size and position of the Window after tracking.

If Tracker breaks lock, there are options. The first is to simply adjust the Window and re-run Tracker. That often does the trick. You are really only using the Window to tell Tracker *what to Track*. Once the Window has been successfully tracked, you can reposition or resize it any way you want. This means you can make it smaller and track just a prominent feature and resize the Window after tracking. Since Tracker uses the difference in Luma and Chroma information, the more the tracked object's Luma and Chroma differs from the background, the better. In fact, you can often improve tracking simply by inserting a Contrast Node before the Window Node and greatly increasing the contrast of the clip to be tracked. When Tracker is finished, you can delete the Node or just turn it off (by clicking on its label).

The second option is to change from the Cloud (area) Tracker to the Point Tracker. Start in the middle of the clip. If you mouse over the Window, you'll get a white crosshair cursor that does nothing–yet. At the bottom left of the Tracker window, select *Add Tracker Point*. That will place a blue crosshair in/near the Window. When you move it, it turns red. Move the red crosshair to a prominent feature (usually an eye, nose, or mouth–or a shiny spot on the forehead). Then run Track Forward from the beginning, reset the play head back to the middle and Track Reverse.

With Resolve 18, the "double arrows" in the Tracker Transport Controls means track forward and backward which saves time.

A third option (switching from Clip to Frame and adding Keyframes) will be covered in the next chapter.

If you can isolate an object (using Window) and follow it (with Tracker), you can add virtually any effect to it (or to the entire frame except for the tracked object if that's what you want).

Intuitive Object Mask

Resolve 18 has an AI engine that can basically do everything you just did with one click. It's called Intuitive Object Mask. Alex Jordon at Learn Color Grading calls it "magical."

First, click on the Face Node, right-click and Reset Node Grade to clear out your previous attempts.

Click on *Magic Mask* (to the right of Tracker). Note that you have two options: to track an "object" or "person." Also, note the eyedropper plus (+) is the highlighted default. *Select Person Mask*. Right below Person Mask is "Toggle Mask Overlay." Turning the overlay on lets you see what the AI thinks you want to isolate. The selected parts will have a reddish tint. Just below Magic Mask are two buttons that allow you to select a Person (head, torso, arms, legs, etc.) or just Features a person might have (face). Select *Features*. There is a pull-down that is rather self-explanatory. Pick *Face*.

Next, simply make a mark on the talent's face with what Resolve 18 calls a "stroke." A checkmark or even a dash will work. With Magic Mask, you do not draw an outline around the face. Simply let Resolve know what in the frame you are referring to by marking it with a stroke.

Since you selected Person Mask > Features > Face, Resolve's AI should have picked only the face. If not, select the negative (-) eyedropper (deselect tool) and identify the selected areas that are not face. You may have to zoom in to better see what's what, but for most purposes, you don't have to include/exclude every single pixel.

Note that the usual Quality controls are available if you need to refine the mask. They work the same way they did with the Qualifier you used for the green screen lesson. Depending on the image, some Quality adjustments won't do anything. However, do experiment with Radius. I find that adjustment particularly useful.

Once you are satisfied that Resolve knows that part of the frame you intend to apply an effect to, Resolve needs to track the entire clip. Since this is one of the most computationally intensive things Resolve does, it can take quite a while, so practicing with short clips (10 seconds) is best. Start the tracker by clicking on the "Track Forward and Backward" double arrows. Depending on your PC's abilities, it may take a minute to start. My PC takes a few *seconds per frame* to process, so it's off to Starbucks.

Fixing Faces

Finally, with the face isolated, at this point, you could adjust color, Sat, or Luma, and those would only affect the face. Try this if you want (you can always reset it). To do that, we are going to add another Node (call it EFX) and connect the alpha channel (blue connector) from the Face Node to the EFX node. Then double click on the EFX Node and try various adjustments and effects on the EFX Node (that way, you won't ruin the Face mask if you reset). And once the Face has been isolated and tracked, you can turn off the Toggle Mask Overlay.

In addition to basic color grading, there are some great tools in Resolve for fixing wrinkles and blemishes. Here I'm referring to the Beauty and Face Refinement plug-ins available in OpenFX (Studio version).

The easiest-to-use facial repair plug-in is modestly called "Beauty." Make sure the EFX Node is selected.

Beauty

Still on the Color page, select Open FX > Resolve FX Refine. Drag *Beauty* to the EFX Node. Beauty has two operating modes (under Advanced Options): Automatic (simpler) and Ultra Beauty (more fiddly). Sticking with Automatic for now, zoom in (or go full-screen). Set Timeline Proxy Resolution to Full and adjust the Amount, Scale, and Global Blend controls to smooth out the most noticeable facial defects. When this effect is applied, some rendering will surely need to take place.

Amount controls how much smoothing you want to achieve, while *Scale* controls the size of the blemishes/defects to be dealt with.

Global Blend works sort of like translucent powder to blend Amount and Scale beautification with the original image. These controls are somewhat interactive, so you'll have to play with them to see how they affect the skin to achieve the blemish reduction/skin smoothing you need.

Like most things in life, too much enhancement is not a good thing. *Beauty* acts sort of like a gauze and can remove lots of detail. That can reduce wrinkles and eliminate blemishes, but too much Beauty removes so much detail it looks like the talent overdid Botox. You'll have to work the controls until you find the Goldilocks settings that work for your onscreen talent. You may want to adjust Saturation, Luma, or Chromas as well. The idea is to slightly buff the image but not remove so many features that they look like an alien (unless that's your objective).

Play the clip and evaluate.

Qualifier

To recap, in the above exercise, you *spatially* isolated a section of the frame and applied an effect to just the isolated portion demarcated by the Magic Mask, whose *changing position* was defined by the Intuitive Object Mask tracker. There are other ways this could be done. One of those other ways is to use the Qualifier. Here we are going to use the Luma and Chroma values of the skin itself to isolate the skin (and hopefully, not much else). It's sort of like Chroma Key, but the skin takes the place of the green screen.

First, select the Face Node and Reset Node Grade. Rest the EFX Node, too.

Highlight

Select the Face Node. Move the play head until the frame in the viewer is representative. Select the Qualifier eyedropper. In this case, the default HSL color model will work just fine. Zoom in on the face and use the eyedropper to select an area of representative skin tone. Turn on the Highlight icon (magic wand at the top left of the Viewer).

The selected area will be highlighted, while everything else will be grayed/black. You can poke around on the face until you find the one spot that more or less captures the entire face without catching (much of) anything else. With this technique, you also get hands and bare shoulders and any other exposed skin, which is often a plus.

Next, under Selection Range, pick the eyedropper plus (+), so you can pick more than one area. Now click in areas that were missed. If you have to reset Qualifier, you'll have to start over with the single eyedropper "picker" in Selection Range.

With some Qualifier uses (such as Chroma Key), you need to select every bit of the target and nothing else, but with faces, if you get almost all of the face almost all of the time (and little else), you should be fine—depending on the effect you intend to apply. Ctrl+Z if you go too far. When you are done, most of the face should be selected by Qualifier.

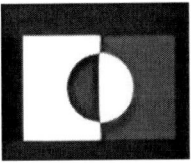

Highlight B&W

Usually, with this effect, you shouldn't need to tweak the selection, but if you need to, here's how: First, select Highlight and B&W (the default in Resolve 18). Under Selection Range, increase Denoise until the image just starts to get blurry, then back off a touch. Adjust the rest of the Matte Finesse controls like you did in the Chroma Key exercise, but keep in mind you don't have to capture every bit of skin (nor will you) or eliminate everything that's not skin (nor will you), so there's no point in overdoing the controls. It's unlikely you'll get completely clean blacks and whites, but that's not necessary here. What you want is to capture as much of the face as practical and very little of the hair or clothing.

Other Highlight

Deselect Highlight. Adjust the Offset Color Wheel to some extreme value to see how Qualifier works. Reset it when you're done.

Drag the Beauty plug-in to the EFX Node. It will only be applied to the areas you selected based solely on their HSL values rather than their position. Adjust Amount, Scale, and Global Blend as in the previous example. You can tweak Luma, Chroma, and Saturation values for just the face as well. A full-screen monitor is helpful (with Timeline Proxy Resolution off so you can see all the detail). Some rendering may need to occur with this effect. Notice that since the Beauty plug-in is active in the areas defined by Qualifier, it automatically follows the face (no spatial tracking required).

While you are at it, you could try out the "Ultra Beauty mode and experiment with what the additional controls do.

Face Refinement

Face Refinement is a much more capable complexion fixer, and once you've tracked the face, Face Refinement performs another tracker (Analyze) to identify eyes, forehead, cheeks, chin, lips, etc. Right-click on the EFX Node, then Reset Node Grade and start over. Reset the Face Node as well. Use any tracking method to identify the face.

When you've identified the face to Resolve, drag the *Face Refinement* effect to the EFX Node and push *Analyze*.

Face Refinement will track back and forth to isolate the specific parts of the face (which may take a while, so use short clips for practice). This effect used to take forever to process, but Resolve 18 reduced the processing time quite a bit. Set the Viewer to *Open FX Overlay* (bottom left tab) and Face Refinement to *Show Overlay* to see this in action.

When it's done, the face will be divided into zones. Unclick *Show Overlay* if it bothers you.

In the Effects tab, click *Settings,* and you'll see zones for lips, eyes, forehead, cheeks, chin, etc. You could change the Contrast, Midtone, Color Boost (Saturation), and Tint for each zone, as well as smooth wrinkles and other imperfections zone by zone. It won't remove double chins, (I've tried) but it can help remove bags under the eyes (to a degree). I won't step you through all of these controls because (a) they are obvious, and (b) you need to see for yourself what they do. Keep in mind that depending on your PC, changes may take a moment to process, even on a single frame and, for a while, it may seem like the controls are doing nothing—be patient.

With Face Refinement, I mostly use Texture's *Amount* and *Scale* to de-emphasize major imperfections. I'll use Color Grading > Shine Removal if the translucent powder stopped working mid-shoot. And I almost always use the Eye Retouching tool to brighten the whites of the eyes a bit. Sometimes I'll add a touch of Eye Light, and occasionally, Eyebag Removal (which doesn't work well with glasses but does sometimes). I didn't count them all, but there must be more than 30 adjustments for this one plug-in. When you've done all you can do to turn them into beautiful people, run the clip and make sure nothing weird happens. If the talent is your boss, now is the time to ask for a raise.

Final Node

After all these tweaks, you may still want to have one Final Node for fine-tuning. Create one and activate Curves. Say you want to ramp up Saturation. Since some parts of the frame may already be at high saturation, simply turning Saturation up may cause the frame to blow out like Technicolor (intense, bilious reds, blues, and greens). But lowering the saturation in the brightest areas can give you the headroom to increase the clip's overall Saturation without such artifacts.

To do that, in the Curves drop-down (defaults to Custom), select the *Sat Vs Lum*. Move the cursor to the center of the white line. There will already be a dot on the far right (brightest highlights). Click to add another one about an inch or two to the left. Grab the center of the line (in between the two points) to lower the saturation a bit for the brightest highlights (while keeping full saturation in the rest). You can now increase Sat without degrading artifacts (*Col Boost* does essentially the same thing).

You can also use this Final Node for any other color corrections you want to make, as well as any fine-tuning needed to blend this clip with the footage on either side.

Exercise Quattro

Picking up from Exercise *Tre*, once you learn how to isolate an area (and track it if it's moving), you can adjust values (Luma, Chroma) as well as add virtually any effect, and that effect will only affect the isolated area (or only the area not isolated if you *Invert*). Here are some examples of how that can be useful:

Blurring a Static Object

Unless you're into doxing, you'll sometimes need to blur personally identifiable information such as credit card numbers, e-mail addresses, or license plates. Shoot a sample clip (10 seconds) of one of those items. The object to be blurred, for this first exercise, should be static.

Create a New Project and import media. Put the clip in the Timeline. White balance, adjust Luma levels, etc.

On the Color page, add a serial Node. Label it "Blur."

To limit an effect to a specific area, select the Power Window control and the Linear (square) or Circle icon depending on the basic shape needed, or draw a free-form shape with the Curve pen. You are not limited to only using one shape and can turn a Window off and on by clicking on its enabling icon (red framed box) in the center panel. (And don't forget that if the Window wireframe disappears, you can get it back using the pulldown at the bottom left of the Viewer.)

Activate Open FX. Left-click and drag Gaussian Blur or Mosaic Blur to the Blur Node. Gaussian Blur is when you want to subtly hide something, while Mosaic Blur is when you want to *show* you are hiding something. Adjust the blur controls appropriately. You can reset the control Settings and start over if you need to.

In addition to blurring an object, you could use this technique to lower the brightness of an area that's too hot. I use this technique frequently if I have my MacBook in a shot because the brushed aluminum case is highly specular. I create a Node with a Window, isolate the Mac, and dial down some of the Gain. It's a matter of esthetic, but I don't like having large, super-bright objects in frame—*distracts from the talent*.

In fact, you can set a Window (or multiple Windows) and apply virtually any effect just to the delineated areas. Now would be a good time to play around with the various Power Window shape options (the slash "/" is particularly useful).

If a Node doesn't play back smoothly, yet for some reason, Resolve doesn't halt and render, you can right-click on a Node, select Node Cache, and turn it On. It's a common practice to "pre-render" difficult Nodes because they usually don't have to be rendered again and again. The rule is to set up problematic Nodes first and Node Cache them. Nodes created downstream can then be rendered without you having to re-render the more complex upstream Nodes again and again.

Blurring a Moving Object with Tracker

As you undoubtedly noticed, this blur was easy to do because the object was static. But physical objects have the unfortunate habit of not staying put. In that case, you can use Tracker or Intuitive Object Mask (*object* mode) to track a moving object and deal with it in a variety of ways.

Say some jerk wearing a tee-shirt with an inappropriate logo walks through frame while you're shooting. You'll need to track the logo before you can apply an effect such as Blur. Shoot a clip that has some moving object in it you'd like to blur—a license plate on a moving car or an actual jerk in a stupid tee-shirt—shouldn't be too hard to find.

Create a New Project and import media. Create a new Blur Node, yada, yada. For this Tracking Blur Node, you'll need to isolate the area using Window, then track it using Tracker before adding the Blur effect. If the clip is long, use the razor blade to cut the clip a few frames before and a few frames after the part you need to track—in effect creating a subclip—and just work with the middle subclip.

Move the play head so the object to be tracked is in frame (completely). Zoom the Viewer in on the object to be blurred (if necessary).

On the Color page, with the Blur Node active, create a Window and isolate the area you are interested in blurring. Window's Curve pen works best in these cases, and to improve the accuracy of Tracker, BMD suggests drawing the wireframe outline fairly tight. But the prefabbed forms may also work for you and are faster (and adjustable).

Apply Tracker as in Exercise *Tre* (meaning turn off any default tracking dimensions you don't really need), then apply the new Intuitive Object Tracker. Then add Effects > Open FX > Mosaic Blur to the Blur Node and adjust values.

Tracker may stop tracking when it can no longer clearly separate an object from the background based on Luma and Chroma differences. In fact, Tracker works best when the object being tracked is easily distinguishable from the background (white car against a dark green forest, for example).

If there's not much difference between the object and the background (Luma/Chroma-wise), Tracker can have problems, and you'll have to use other methods such as adjusting the wireframe's size and shape, and retracking. Exercise *Tre* should have prepped you for making adjustments to the Cloud Tracker or using the Point Tracker. You could also use Intuitive Object Mask's "Object" setting.

Tracker works better if the footage is high res (4K vs. HD). Certain file types (H.264/265) seem to be more problematic than others unless they've been "optimized." And although I said to turn off extraneous dimensions because they add to processing time, you may need to add back those dimensions if the object actually moves in those planes. For example, if you are only tracking in Pan and Tilt and the object moves toward or away from you, add Zoom and re-Track. (Basically duplicating what moves a camera would make when tracking the object.)

If an object has a prominent feature (a logo or other clearly distinguishing characteristic), put the wireframe around just that feature, track just that feature, and you'll probably get better tracking results. Once tracked, you can change the shape of the wireframe to include the entire object of interest. I find this method works almost all the time (assuming the object has a conspicuous attribute). Point Tracker (previous Chapter) does basically the same thing, only better.

And if you simply can't get Tracker to work properly despite best efforts (and, admittedly, it is finicky), you may have to watch a few YT videos to get comfortable with it. Most YT videos are simply demos where you can see what an expert can do with it rather than a tutorial that will teach you how, but there are a few of those, and they are worth watching if you are stuck.

Keyframes

Keyframes are basically Post-It notes that tell tools like Tracker what to do at a given video frame. Keyframes are used by several tools, including Tracker.

Tracker Keyframe

If you have trouble getting Tracker to track, play the clip and stop at the point where Tracker loses track. Change Tracker from Clip to *Frame*, and manually nudge the position of the Window. This action will automatically generate a Keyframe. Move the play head a few frames (using the arrow keys), then readjust the wireframe. Wash, rinse, repeat. This method can be tedious, but it does work if all else fails. That said, the new Intuitive Object Mask is probably your best fallback if the object proves difficult to track—it's easier to use, more accurate, and, well, "intuitive." Also, the BMD Reference Manual has a very detailed explanation with nicely illustrated instructions (Motion Tracking Windows), and you can get to it from Resolve's Help menu.

Tracking is an important tool to master because once you can track a moving object, you can apply all kinds of effects (works with text, too).

Patch Replacer

Out, out damned spot! Sometimes, simply blurring something you want to hide may draw more attention to it than you intend. In that case, obscuring the offending image/object may not be enough—you want it outta there. For that, you can use the Patch Replacer tool to cover a *static* offending area with a patch you've copied from a matching area. It's very simple to use.

Create a Patch Node. Using a clip with an unwanted object, Apply Patch Replacer from Open FX > Resolve FX Revival > *Patch Replacer*.

You'll get two ovals–input and output. Place the input oval over a clean area you want to clone from. Place the output oval over the spot to be covered up. Adjust the size of the output oval as necessary (the input oval's size will follow the output's). There are also finesse controls in the Inspector to fine-tune should you need to. To remove more than one spot, create additional Nodes (one for each spot).

Object Removal

If the offending image is moving, you can use the *Object Removal* tool to completely remove an object from a clip. Of course, you can't just "remove" something without creating an obvious hole, but Object Removal can create a "clean" background plate to fill in the hole.

To learn how to use Object Removal, shoot a *locked-down shot* of a scene where a car enters and exits frame. Create a New Project, import media and all that. Cut the clip into three pieces as described earlier (so that the processing computations needed to pull this off are only applied to the middle clip where all the offending action is). Select the middle subclip (the one with the moving object to be excised). Add a Node and label it "Remove Car."

Intuitive Object Mask in Resolve 18 is easier to use to track a moving object than Tracker, but you can use either Tracker or Intuitive Object Mask to track objects.

If you are using Tracker, select a Power Window, and draw a wireframe (Curve pen) around the car. BMD's documentation advises drawing the wireframe pretty tight. Also, BMD recommends "some" *Softness* (bottom right of Window).

Switch to Tracker. Place the play head in the middle of the clip and Track Forward. Then move the play head back to the middle (there will be a keyframe dot where you started from) and Track Reverse. Scrub the clip and make sure Tracker is tracking accurately. If not, make the necessary fixes. And make sure the wireframe is always outside of the object to be removed (adjust if necessary).

At this point, whether you used Tracker or Intuitive Object Mask or some other method to identify the moving object, the next step is to remove it. Select Open FX, scroll down to Resolve FX Revival, and drag *Object Removal* to the Remove Car Node. BMD's documentation says to right-click on the Node and enable *Use OFX Alpha*, but in Resolve 18, that's the default.

Under Analysis, select *Assume No Motion* (if the camera is, in fact, locked down). Assume No Motion refers to the *camera's* motion, not the object's. If Resolve doesn't have to track a camera move *and* a moving object at the same time, it doesn't have to "analyze" the scene.

But if the camera is also panning, tilting, or zooming while the object is moving, you'll need to use *Scene Analysis*. This process can take a lot of time, depending on the complexity of the camera's motion and the length of the clip–meaning you can probably make it to Starbucks and back. Otherwise, select Assume No Motion and skip Scene Analysis.

At this point, Object Removal is going to remove the wireframed area and replace it with background. If it can't do this on its own, you'll get a gray matte that will need your attention. Under Render > Search Range, increase the value (slowly) until the gray matte goes away (but don't go any further than you have to). Scrub from end to end. If additional gray matte areas appear, click on Build Clean Plate and play the clip.

You may see some gremlins. How to fix them depends on what's causing them. If there are shadows you didn't catch with the wireframe, you'll have to expand the wireframe until it does. (Using *Show Mask Overlay* at this point will help you figure that out.)

You can also try changing Blend Mode to Adaptive. Some people say that works, but I've never gotten it to (I get anomalous glitches).

Moving the play head to the end or beginning (or just some random location) and building another clean plate using a different frame actually does seem to help.

Success in removing a moving object is highly dependent on the source footage and usually requires fiddling (and practice). When you're done, go full-screen and play the clip (will need to be rendered). Pretty damn cool, no? You can apply virtually any Open FX effect to a static or moving object using this method.

Surface Tracker

New for Resolve 18 is Surface Tracker (a/k/a/ Surface *Mesh* Tracking). The previous trackers and masks tracked 3D objects. Surface Tracker tracks *surfaces* (paper, fabric, skin). Most of the examples I've seen on YouTube use Surface Tracker to place a tattoo on someone's arm or cheek. How often that's necessary is anybody's guess. But it is a neat effect, and it's new with Resolve 18. Try it, and one day you may find a use for it.

How would you use it? A former client had three distinct brands of the same product. Sometimes we shot three videos with the talent wearing a different branded shirt each time. Other times we shot a single unbranded video and added a "bug" (a tiny persistent logo). With Surface Tracker, we could have added each brand's logo to the talent's shirt, shooting only one version.

To use Surface Tracker (an Open FX effect on the Color page), you draw an outline (fairly precisely) of an area (only that part you intend to track). A "mesh" is created that maps the surface (and you can adjust the mesh if necessary). The area covered by the mesh can be replaced with another image (usually a graphic or still) even if the surface below it moves, ripples, undulates, etc.

To learn how Surface Tracker works, shoot (or otherwise obtain) a 10-second clip of someone wearing a polo or tee-shirt, for example. Also, obtain a logo (jpeg) of some kind. Import those into the Master Bin.

Create a serial Node and call it Surface Tracking. Make sure it's green output is connected to the video input (right) in the Nodes area.

Open Effects, find Surface Tracker and drag it to the Surface Tracking Node.

The default tool is *Bounds*. Click on an area on the shirt where you want the logo to be and create a closed box (defining all the necessary points). The more points you select, the more accurate Surface Tracker will be. Click *Mesh* to see what Surface Tracker's AI is seeing. You can increase the "Point Number Limit" for a finer Mesh if necessary. Or you can create Bounds with lots more points.

Then click *Track*. In some cases, you'll need to change *Quality* from Faster to Better, but Better takes longer to process. For most uses, you can simply click on the double arrows and track in both directions. Surface Tracker does lots of mathematical calculations, each Surface Track takes two passes, so the process can be a long one, and it's best to practice with short clips. If the Surface Tracker fails to track accurately, try redrawing it. Also, adjusting the Motion Range and Mesh Rigidity may be necessary.

When Tracking is done, click on *Result* to see the result.

Now, right click on the Surface Tracking Node and select *Add OFX Input*. This creates another input to the Surface Tracking Node.

Now grab the graphic (logo) and drag it directly to the Nodes area. An External Matte will be created. Connect the green video output of the Ext. Matte to the green input of the Surface Tracking Node.

Note that under the Compositing tab in Result, you can affect the Composite Type as well as the opacity. And Hide makes all the behind-the-scenes disappear. Way cool, no?

Smartphone Video

Smartphones now produce such high-quality video that there are many YouTube videos shot entirely on smartphones. Even if you are using a camcorder or DSLR, you will probably find it necessary at some point to incorporate smartphone video.

How to get the video off your smartphone and into your computer depends on which brand of smartphone you have. With the iPhone, you can transfer footage from the phone to iCloud. The resulting file will be a .mov that can be imported just like any other video file. Of course, the frame rate and resolution may not match your current Timeline, but Resolve can handle that.

For reasons I do not understand, a majority of smartphone footage is shot vertically. As you know, the world, your eyes, and every movie and TV show ever made were all shot horizontally. If you have vertical smartphone footage, the first thing you'll want to do is change it to a horizontal clip. The most common way of doing that is with heavily blurred sidebars on either side of the vertical footage.

For this exercise, shoot some smartphone footage. It pains me to say, "shoot it vertically," but that's the drill. Transfer the footage to your computer using the methods required by your phone and carrier.

There are several ways to turn a vertical smartphone video into a horizontal one. Here's one way:

Click on the smartphone clip so that it's in the Source Viewer. From the Source Viewer screen, drag it to the Video 1 track. (Just drag the video at this point by pulling from the film frame near the bottom of the Source Viewer.) Resolve may ask you if the Timeline frame rate should be changed to match the phone's video. In some cases, that's

OK, but most of the time, you are working with a frame rate that you'll want to keep.

Once imported, it's probably a good idea to Optimize smartphone footage. My iPhone footage wants to be rendered immediately in any case.

First, we'll create the blurry sidebars. On the Edit page, open the Inspector and zoom in (X-axis) on the video until it's full-screen.

Open the Effects Library > Open FX and drag *Gaussian Blur* onto the Video 1 clip. Adjust the blur in the Effects > Open FX Inspector as appropriate (usually quite a bit).

Drag the same smartphone clip (including the audio if you need it) to Video 2. You may want to zoom in a bit (Inspector > Transform > Zoom) to fatten the image or even move around in the image (X,Y) until it shows what you want.

On the Color page, select the Video 1 clip (the sidebars), create a new Node, and reduce the Luma for Gain and Gamma of the sidebars if you like. Reducing Saturation is sometimes appropriate as well.

Creating YouTube Thumbnails

On YouTube, it's the thumbnail that actually gets people to watch your video, and if it isn't compelling, potential viewers won't click on it. Resolve has tools for creating thumbnails—sort of. Adobe Photoshop, Affinity Designer (better and cheaper), or some similar photo editor would be a better choice (mainly because of the ease of changing the resolution and other attributes). But you can use Resolve to make YouTube thumbnails. and many do.

Grab Still

You could search your video until you found an appropriate image for a thumbnail, but IMO you'd be better off shooting one specifically for the thumbnail with a *still camera*. Most YouTube thumbnails have exaggerated facial expressions like "surprise," or "wonder," or maybe just an over-the-top smile. Unless your talent is on something, you're probably not going to find a still you like by rummaging through your footage. It's usually better to shoot a few seconds of video at the end of a take and have the talent mug for the camera. But because of Resolve's clumsy way of grabbing stills, it will be much easier just to shoot a still for this purpose with your DSLR. But if you absolutely need to strip a still from a video track, here goes:

An often-used method involves *Grab Still*. On the Color page, select the clip in the Timeline you want to grab a still from. Scrub until you find the exact frame you want. *In the viewer*, right-click on the image and *Grab Still*. A "still frame" will pop up in the Gallery. The still will be at the same resolution as the Timeline. When you hover over the "still," the thumbnail view changes but the underlying image does not.

Right-click on the still and choose Export (or Export with LUT if appropriate) and select the format you want (usually PNG or JPEG). Change the storage location and give the still a name. A good practice is to associate the still's name with the clip's name.

While the long-established rules for video/graphic design don't seem to be consistently applied to YouTube thumbnails, there are some very good examples to follow. Some of the best thumbnails on YouTube (IMO) are done by Sara Dietschy (rhymes with peachy). In fact, everything about her YouTube presence is there to be emulated. Fittingly, Dietschy uses DaVinci Resolve. Check out her Resolve videos because she includes links to a list of Resolve shortcuts for both Windows and Mac that you can download from her Google drive or Dropbox. Even though she puts out several videos each week, her thumbnails are consistently top-notch and eye-catching.

Add Text from the Effects Generator as necessary. When you are done, Grab Still of the final product and export. Done this way, the resulting thumbnail should meet all of YouTube's specs.

Adjustment Clip

Adjustment Clip is a blank clip (contains no video or audio) that you can apply transforms (zooms, crops) –even effects and grade Nodes. Then you can copy the Adjustment Clip and apply the same "adjustments" to multiple clips—saves tons of time.

Start a new project and import a clip. Drag the clip to Video 1. From the Edit page, go to Effects Library > Toolbox > Effects > *Adjustment Clip*.

Basically, grab Adjustment Clip and put it in the Timeline over a clip you want to "adjust." Go to the Color page, click on the Adjustment Clip, and make all the adjustments you want. These will be stored in the Adjustment Clip rather than in the video clip itself. Once done, you can put any clip you want under the Adjustment Clip and get the identical "adjustment." This is often easier, faster, and better than copying a grade from one clip to another (and it's reusable).

Rules

Of course, you knew there would be rules, didn't you? First, the Adjustment Clip has to be the top-most Track. It also affects *all* of the Tracks below it. Furthermore, the Adjustment Clip only affects those sections of the lower Track(s) that the Adjustment Clip actually covers (usually that's 100% but not always). You can also fade the Adjustment Clip in and out. Basically, anything you can do to a clip, you can do to an Adjustment Clip.

I have one Adjustment Clip that took a while to get exactly right. It's actually several Open FX effects and grade adjustments that make an HD clip look like VHS (also cropped to SD). It's from a recipe I found on the dark web IIRC, and it's a PITA to reproduce. Instead, since I knew I would want to reuse that effect at some point, I created it as an Adjustment Clip and copied it to a "project" that stores just my "Adjustment Clips." I can copy and paste that Adjustment Clip into a new project whenever I need to. More importantly, Adjustment Clips can be changed—even copied ones. I can make the VHS effect B&W, for example.

Look or Listen

The guys who taught me the editing business had a unique way of reviewing the final edit. Their point of view was that television and film were basically an amalgam of silent pictures plus radio. In silent films, you had to be able to tell the story with just pictures (and some graphics). In radio, you had to be able to tell the story with just sound.

With television, if you turn the sound off, and the video alone more or less tells the whole story—you can follow it, and it makes visual sense—the video is pulling its own weight. Then, if you listen to just the audio with the video off to hear what the audio track sounds like by itself—and it too tells a coherent story– then the combination will work that much better.

People see with their eyes and listen with their ears, and these two separate tracks of information are processed by separate areas of the brain. Sometimes we use audio as a crutch to cover what we don't have on video and vice versa. Sometimes that's a necessity. But it's a weak practice to be avoided if possible. If you view the video silently, you can see far more things than when the audio accompanies it. Same is true with the audio—you can hear more if you see less. And if the video and audio work standing alone, when combined, you'll have something that is greater than the sum of the parts.

Cuts that seem to work OK with the audio on expose their shortcomings with the sound off. This is particularly true with titles and graphics. If the video doesn't tell the story with the audio off, it needs fixing—titles, graphics, B-roll—something.

When the video is fixed, and the audio is laid back in, it's a whole different thing. Fumbled edits and auditory speed bumps tend to get papered over by the visuals. Key points that are only conveyed aurally may need titles added so that they are not missed. Identify and fix any weakness in the audio track, then turn the video back on. The two tracks will now reinforce each other rather than cover up each other's faults.

I once edited a promo video that was to be played on monitors in bars. The client sent us some examples of what their competitors were doing and wanted us to use a similar approach. Nice videos, really. Smooth shots, upbeat music, polished script, nice vocal track.

The problem was, bars are noisy There's often competing music. No one is paying attention. In my view, the audio track didn't really matter because, when we checked, *none* of the bars actually played these videos with the sound on. So we went 100% visual: flash cuts, pretty girls, splashes of color, plenty of bold text, and, of course, animation.

We put a scratch track on it just in case, but I don't think anyone ever heard it.

Parting Shot

At this point, if you've done the four exercises (more than once), you should be able to figure out how to do most anything you need to in Resolve, as well as follow the more advanced YouTube videos.

At this point, you are ready to go beyond the basics. To help you expand your horizons, there are additional exercises in the chapters on Fusion and Fairlight. The Resources chapter has a curated list of YouTube channels, tutorials, guides, manuals, and other information that will help you take it to the next level. Review the next several chapters to familiarize yourself with what's going on in the Pages. Every Resolve Page has its hidden secrets. It's up to you to discover them.

Pages

The Media Page

The Media page is usually the first stop in Resolve because it's the place where your footage gets discovered/imported/ingested. You can also import media from any Resolve page (except Deliver). However, the Media page does have more tools and features for importing media (such as a browser you can use to look for files anywhere on your computer) and systems to help you keep track of media for large projects.

Since you have some familiarity by now with all the Resolve pages, the rest of the chapters are going to focus on features not covered in the exercises and hidden gems that may have gotten short shrift.

> **CAVEAT**: Although Resolve has sophisticated tools to manage media (Smart Bins, Power Bins, and more), I recommend you do not bother with these for short-form projects. These file management tools have their own learning curve and are really designed for complex, long-form projects with lots and lots of pieces. I realize that many Resolve tutorials spend a lot of time helping you learn these file management systems, but I'm afraid you'll spend more time figuring out how to use them than you would if you just dumped your footage into a single Master Bin, created a few dumb Bins and be done with it. Of course, they are there if you need them. In most cases, a single Master Bin and color coding clips will serve you just as well.

Media Page

From the Media page, you can organize clips, view them, scrub them, select I&O points (create subclips), add notes (metadata), and a bunch of other stuff. The very first thing I do on the Media page is kill the 16-track audio mixer (click *Audio*). Why anyone would think that would be useful totally escapes me. I can think of only one camera that outputs 16 track audio. (OK, I lied. I can't think of any camera that does that.)

Media Storage

Starting from the top left, Media Storage gives you access to all the drives on your machine. The list is prioritized based on the drives you previously listed in Preferences > Media Storage > Media Locations. Because Media Storage takes up a lot of screen space, turn off when not needed.

Browser

Sidebar

Below the Expander icon (down arrow, top left) is the Sidebar icon. This opens or closes the Browser. The default is on. If you turn it off, you can still see the thumbnails of the clips of the selected media folder.

You can use the Browser to search for footage on any drive/folder on your machine. The list is prioritized based on global Preferences. As you will recall, the first drive on the list is your cache drive.

Favorites

You can select any folder, right-click and make it a Favorite. Favorites will then be available without you having to search for them again and again. If you have intros, outros, stingers, logos, and bumpers for your channel, put all those in one folder and make it a Favorite (Add Folder to Favorites).

Copying Clips

The Browser brings up the root of all of the drives currently attached to your machine. This means that you can access footage directly from an SD card (via a USB card reader). However, *when you eject that drive the footage will disappear,* so you still need to *physically copy* that media to a drive on your machine. The best way to copy from an SD card, external drive, or camera is the Clone Tool.

Clone Tool

To copy an entire folder (i.e. an SD card) to your project's media folder, first create a destination folder on your media storage drive. Once it's there, you can clone files to it. Since I often forget to create necessary folders in advance, I created a temporary Media Import folder on my media drive and usually clone all media to that folder temporarily. Once all the files are cloned, I create a project media folder and move all the files to that.

Select Clone Tool, but first, open the middle three dots menu and check *Preserve Folder Name* if that's important to you. Checksum Type forces Clone Tool to verify the integrity of what it has copied (a very good idea). You should still make a backup copy of this folder on another drive if you are the belt-and-suspenders type or if the footage is irreplaceable.

Click on *Add Job. Using the Media Browser,* navigate to the folder that contains your project's footage. This could be an SD card, an external drive, or even the camera itself. Drag the folder to Source (or right-click on the source folder and *Set as Clone Source). Using the Media Browser,* find the Media Import folder and drag that to *Destination* (or right-click on the destination folder and *Set as Destination).* Then click *Clone.* Resolve will copy all the files from the Source folder to the Destination folder. You can't use Clone Tool to copy individual clips, just folders, but Clone Tool saves lots of steps, and the ability copy footage directly from a camera or other external storage device is nice.

Media Management

This is one of the most useful additions to the Media page (File > *Media Management).* Like Optimized and Proxy Media, Media Management can copy media or transcode it from what you have to what you want and with many encoding options. For this reason, I don't think you need any additional third-party software to transcode your camera footage to something Resolve likes better.

You can copy or transcode all the clips in the project or individual clips. However, clip-level transcoding can only be done on clips that are already in a Bin. If you do it this way, click on *Relink to new files* so the transcoded version will be substituted. (You can verify this was done by looking at the file type in Metadata.)

Video brings up the codec/encoding parameters. The suggested codec for smoother operation is DNxHD for HD or DNxHR for 4K (ProRes on a Mac). Setting the encoder to the hardware encoder on your graphics card (if that option is available) will speed things up. As to Quality, set that high: 40,000 at least; 80,000 is also a good choice.

> **CAVEAT**: Unlike Optimized Media and Proxy Media, when you transcode using Media Management, you will not be creating just an edit proxy where the original file will be used at the time of delivery. Instead, Media Management transcodes the file and *you'll use that file for editing and delivery*. So setting the encoding parameters matters for more than just how it looks in the edit. For that reason, use Optimized Media or Proxy Media if all you want to do is create an easier-to-edit proxy file but still want to use the original source footage at final.

Media Management has several copying tools that help you manage your footage. You can grab all of the clips and move them to an archive folder, for example. Or you can specify "Used media" and copy only those clips that actually are used in your project. (Add frame handles so you can later add a transition if you need to.)

Media Storage, Media Pool, Master Bin

The clips in Media Storage are available to your project but are not accessible until you drag them to the Media Pool (which then becomes the project's Master Bin).

Smart Bins

If you are working on yet another *Star Wars* remake, Smart Bins will prove useful. For less ambitious projects, just working out of the one Master Bin will probably be fine. If you have footage shot on different cameras, at different frame rates, or of different subjects, you could create Bins for each media type. Smart Bins are populated based on rules you establish. And as BMD's documentation says, it's a fast way for *your assistant* to organize content. If you have such an assistant, I'm sure they already know how to use Smart Bins. If it's just you, why bother?

There are also *Power Bins* which are hidden by default. Good. Bins are just folders, and if you need to organize your stuff, right-click in the Master Bin and just add additional Bins. They won't be Power or Smart, but they'll do. (If you want to see them, go to View > *Show Power Bins*.)

Viewing Clips

Both the Browser and the Bins have viewer controls that let you view clips as icons or as a list. Using the controls immediately above the thumbnails, you can change the size of the thumbnails, sort by name, date, or other metadata. You can also search by filename and other attributes.

If you are viewing clips as thumbnails, you can hover over the thumbnail and scrub to select a frame that's representative of what's in the clip.

Source Viewer

Ctrl+F switches to a full-screen Viewer (and back). While in this mode, you can't see the transport controls, but you can still play and stop using the spacebar. If you move the mouse, you'll get a play/stop control and an audio kill switch. If you set I&O points, you can adjust them if necessary by moving the dots. As mentioned, if you are setting I&O points with the idea of creating subclips, go big.

Metadata Panel

There is also a Metadata panel(top right) that shows all the clip's metadata such as the file type, frame rate, date shot, and other facts.

Behind the Three Dots

Behind the Source Viewer's three dots is a *Live Media Preview* that transfers the thumbnail image to the Source Viewer (with audio) whenever you are scrubbing a thumbnail. Very handy and should be the default.

Inspector

With Resolve 18, the Inspector was upgraded and now lets you Transform (zoom, scale, move), Crop, and Retime clips directly on the Media page. You can also set the Volume, Pan, change Pitch, Speed, and even EQ from here.

Subclips

The longer the clip, the longer the render. It's a good practice to review clips early on and at least cut out the parts you know you will never need. Unless you are very parsimonious when shooting, cutting out extraneous material prior to starting the edit is essential.

Resolve's way of paring down is *subclips*. To create a subclip, select a clip, set I&O points in the Source Viewer, and drag the result from the Source Viewer to a *Bin*. You'll get only what you asked for. The term "subclip" will be appended to the filename, but the names are not incremented in any way, so it's a good idea to give them a new Reel Name or at least number them if you are pulling more than one subclip from a clip.

> **CAVEAT**: If you create subclips, you will have problems later if you try to expand (lengthen) them in the Timeline. It can be done, but it's a PITA. So leave a little extra on each end. Another way is to wait until you are on the Edit page, make your I&O selections there, then drag the resulting clip *directly to the Timeline*. That doesn't create an official "subclip," but done that way, you can expand its duration (up to the entire clip it came from) just using the Resize tool (grab one of the ends of a clip).

Optimized Media

If you need to transcode your files using Optimized Media, you can do that one-off in the Timeline (or in batches from a Bin). Select clips in a Bin that need to be optimized, right-click, and *Generate Optimized Media*. Then, figure out something productive to do with the rest of your day (depends on the settings you established in Project Settings). As of this writing, Resolve doesn't currently show you which clips have been optimized and which haven't. It's a good practice to color code optimized clips until BMD fixes that. Purple is nice.

Proxy Media

Likewise, you can generate Proxy Media from the Media page. The same rules apply. I am finding that Proxy Media generates faster. Whether Proxy Media files are easier to handle than Optimized Media IDK (should be the same or better). However, with Resolve 18, it appears that Proxy Media has become the go-to.

Pre-Editing

The first step in editing is assembling all the pieces in the Master Bin (or Bins). The next step is scrubbing the clips and selecting the I&O points for the pieces you need, making subclips, renaming and sorting, optimizing, and color coding.

If you color code all your clips, you can sort them by color. If you approach a project this way, by the time you are ready to begin the actual edit, you should have all your media imported, reviewed, scrubbed, and trimmed using I&O points, and in the edit you can concentrate on storytelling, effects, grading, or what have you.

The Cut Page

The Cut page, which was added to Resolve beginning with release 16 and somewhat improved in 17, is really aimed at TV news departments that need a fast, efficient way of cutting tape and getting it on air quickly without a lot of falderal. The idea is that the Cut page has every tool you really need—media input, editing, graphics, audio, and media output that will let you create short-form videos from start to finish quickly using only the Cut page.

Actually, I didn't intend to say much about the Cut page because I thought it was just a stripped-down version of the Edit page—sort of a "Resolve for Dummies." It's not that. In some ways it's quite innovative if your workflow happens to match the Cut page's capabilities and idiosyncrasies. But the hype to the contrary, it's not a "starter" step for the Edit page. In fact, it's a lot like Final Cut X (which may explain my animosity).

With the Cut page, Resolve has sort of reinvented the edit interface, and it's just possible that the Cut page is going to be the go-to for many projects particularly if they are straightforward, short, have a repetitive workflow, and need to be performed very quickly. And because Cut page is more efficient in its use of limited real estate, it's ideal for Resolve on a laptop.

However, if you are new to video editing I would strongly urge you to stay away from the Cut page until you have some more conventional editing under your belt. It's really designed for the hyper-specific workflow of TV news, and unless that's what you're doing, the rest of Resolve's tools are much better. If you are new to Resolve, it's the wrong starting point because it is nothing like how the rest of Resolve works.

When it was released at NAB in the spring of 2019, there was a lot of fanfare. BMD touted it as a "one size fits all" solution for high-volume editing operations such as TV newsrooms. There was an initial flurry of blog posts and demo videos posted to YouTube, and then, that cricket sound. The Cut page has a lot of promise, but even with the release of Resolve 18, it also has a lot of problems. The comments on Blackmagic's own Forum have been brutal. So, again, if you are just starting out editing with Resolve, I don't recommend the Cut page at all.

Blackmagic's *Beginner's Guide* takes the opposite view and actually focuses on the Cut page as a beginner's entre to Resolve. That seems to be more of a marketing choice than an objective decision. I think it's exactly the wrong approach. Some of the Cut page features still don't work like they should, and a few are still missing. In my view, it should be the last part of Resolve a new user gets exposed to, not the first. And even then, it's designed for a very specific workflow that most new users are unlikely to be doing.

BMD's *New Features Guide* also spends a lot of time on the Cut page. They seem to really want to make the Cut page part of the standard editing workflow. Of course, a small (and inexpensive) "Speed Editor" hardware controller has been released specifically for the Cut page. Bing.

If you do decide you want to give it a shot, watch several *current* YouTube videos about the Cut page first—written explanations about how the edit functions work really don't do it justice and also don't fully expose you to the obstacles.

That said, there aren't a lot of current videos about the Cut page. Several fan-base videos dropped with the launch of the Cut page. So there are a lot of "demos" out there touting the features of the Cut page but very few honest-to-god tutorials—even more than a year later. (The exceptions are the Cut page tutorial by *Learn Color Grading* and Darren Mostyn's Cut page explainer–both on YouTube.)

> **UPDATE:** The above review is a little harsh, and after working with the Cut page a bit, I have modified my views and moderated my tone. BMD has a "Speed Editor" control panel especially designed for the Cut page ($295 from B&H) that makes the Cut page a much more viable option even for casual users. Using a BMD control panel with Resolve is a completely different experience and reinforces the point that BMD is a hardware company, and Resolve is, at the end of the day, really designed to work with BMD control panels.

Spood Editor

One Viewer to Find Them All

Because you only have one Viewer on the Cut page, you get to decide what it shows. The first button makes it a Source Clip Viewer—that's the default. The third button turns it into a Timeline Viewer. Immediately to the right of the Timecode display is a pull down that lets you change the resolution of the project without having to change it in Project Settings.

Source Tape

Source Tape

In between the Source Clip and Timeline icons is "Source Tape." Source Tape lets you view all the clips as if they were just one big clip.

A long time ago at a TV station not too far away, we shot the entire story—interviews, B-roll, cutaways, close ups, crowd scenes, stills, *everything*—on a single tape cartridge. Before that, we did pretty much the same thing on a single roll of 16mm news film. When it was possible, we shot the material in the order we intended to use it. That made it easy to get a story on air quickly because speed was the demon that drove us all. Source Tape is the same concept. It really only works for smallish projects. Too many clips would choke it. But for a small number of clips, it could be handy.

Note that the order of the clips in Source Tape is based on the order of the Sort. Since you can Sort by name, numbering your clips in the order you expect to use them and sorting by name will build a Source Tape in that order. If you are fairly judicious in selecting I & O points in each clip (and create subclips), and you name them so they can be sorted in order, by selecting Source Tape you have basically created a rough cut of the entire project. By playing the rough cut, you'll have a better idea of what to take out and what further edits need to be made than by building the Timeline by adding one clip at a time.

In Source Tape, using the up and down arrow keys jumps the play head from clip to clip. And grabbing video from the Viewer pulls the entire Source Tape (all the clips) into the Timeline.

Now with Two Timelines

To keep from having to zoom in and zoom out of the Timeline constantly in order to have either a bird's eye view or a surgeon's accuracy, the Cut page gives you two Timelines at the same time, all the time. The lower one is similar to a zoomed in Timeline on the Edit page. The upper one shows the Timeline zoomed out all the way, so you can see your entire project from end to end. You can even drag clips from the upper Timeline to the lower Timeline and vice versa. This is another area where color-coding clips by type would come in handy.

Not having to zoom in and out constantly to see the entire project or cut at frame level is a big advantage. While I really like this approach, there's no reason the Edit page couldn't be equipped with two Timelines as well (optional).

Lock Play Head

On the Cut page, you can operate with the play head locked (you move the clips) or "free" (you move the play head). (Look for the "T" icons with the lock symbols.) Most NLE operate with the play head "free" and the play head moves down the Timeline. With the play head locked, the Timeline moves past the fixed play head. I'm guessing the TV newsroom editor they're competing with also has this feature. However, you already have two Timelines on the Cut page, so you can lock the lower one (the upper one is always free) and that does add some value.

Vive la Différence

There are differences between Cut and Edit that you need to be aware of—some very major. One of the most obvious differences is that you only get one Viewer on the Cut page, but honestly, most of the time that's all you need. We only had one viewer on the Moviola and flatbeds now that I think about it.

But even with one monitor, with everything that's happening on that one page, the screen can get a bit cluttered, so unless you have lots of clips, grab the right side of the Media Pool and shrink it. Grab the top of the upper Timeline to shrink it too, which enlarges the Viewer (this doesn't work on the Mac AFAIK).

Cut Page Panel Controls

Import Media

On the top-most control ribbon, starting from the left, is a sidebar icon that actually opens a mini browser.

The next button is new and is not on the Edit page—*Import Media*. Handy. Should be on the Edit page, too, IMO. Next to it, the Import Media Folder can import an entire folder of media with one click.

Thumbnail view

Then there are the views. The standard view is Thumbnail. When selected, Thumbnail View lets you scrub the thumbnail while it's still in the Media Pool, yet you can see it in the Viewer. You can hear it, too. To the left is the new Metadata View that lets you view clips including key metadata fields.

Filmstrip view

In the Media Pool, you also get a button that displays clips—wait for it—*as clips*. Try it. If you just hover over a clip and move left-right, you can see the clip "play." If you click on a clip, you can scrub it in the Viewer. It's officially known as *Strip View* (as in *filmstrip*). Again, that's a feature that the Edit page could use as well.

That's followed by the List View. Like the rest of Resolve, you can Search the clips by keyword. When it finds one, the rest go away. You can get them back by clicking on the "X" in the Search window. Works just like Search on the Edit page.

Transport Controls

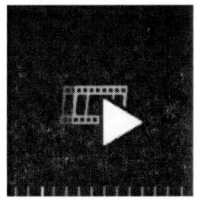
Fast Review

The controls for the Viewer are typical except for the leftmost two. The "play/film frame" icon on the bottom left of the viewer –*Fast Review*– is simply not usable IMO.

Supposedly, it lets you play through clips quickly so you can review lots of material fast. It doesn't really work. The play back speed is not adjustable (by you) as it *depends on the length of the clip*. A short clip plays back in nearly real-time but slightly faster. A long clip plays back in very fast forward. But there's no pitch control (and no Master fader) so I find it unusable for scanning SOT. Maybe it would be useful for going through a lot of B-roll but not much else IMO.

The multi-slider button under the Viewer pops up several icons (Timeline Viewer mode only). The first one lets you move the image around. You can resize with this tool as well. Next is Crop which does just what it says. The musical icon activates a *clip-level* volume control.

Master Fader?

While there is an audio volume display for clips on the Cut page, there are no Master or Track faders AFAIK. The volume control underneath the Viewer has to be set for *each individual clip*. You can work around not having a Master fader, but again, it's something I thought they'd get to at some point, but they didn't with Resolve 18 Even at the BMD live demo at NAB, they had to sneakily switch to the Edit page to adjust the Master volume—which negates the "all-in-one" concept of the Cut page I would think. Not a big deal, but further evidence that the Cut page was released before it was fully baked.

AI Is Going to Save Us

Moving down to center screen, at the far left is the *Boring Detector* (zzzZ icon). When I first heard about this, I thought it was a joke. But Resolve now has an AI-like feature that can scan your video and tell if it's boring or not. Really. You don't need a boss anymore.

Actually, it's a lot simpler than that, and it could possibly be handy in a news room where there's just not a enough time to review an edited news package before it's uplinked via satellite to the entire world.

How It Works: A news room might have a standard that says a talking head in a news package can't be longer than 45 seconds without management approval. The Boring Detector scans the Timeline for such policy violations. You can set that parameter to however many seconds management tells you to. BTW, once you activate the Boring Detector, it's persistent. It stays active until management tells you you can turn it off.

Cut, Dissolve, Smooth Dissolve

Unlike the rest of Resolve, to Cut, Dissolve, or Smooth Dissolve, you place the play head where you want a transition and click one of three buttons. This doesn't seem to offer any advantage over the normal way you would do it on the Edit page, but I suspect if you were using BMD's new Speed Editor control panel for this that would indeed speed things up a bit.

CAVEAT: The thing about Cut page that confuses most new users is that Track 1's behavior is very different from the rest of the tracks. When you adjust a clip or delete parts or all of a clip in Track 1, the Cut page will automatically ripple to delete any gaps. Cut page doesn't do that for the rest of the tracks (they operate just like on the Edit page). But Track 1 behaves differently, and that can be confusing if you are not aware of this different behavior.

The Edit Page

If, in some distant galaxy, far, far away, a Droid asks you to "Shut down all garbage mashers on the Detention Level," that's probably just an on/off button, right? No problem—*as long as you can find the damned button.*

The same thing can be said of learning Resolve. The most difficult part is often the *where*, not the why or even the how. So knowing where the buttons are and what they do is an important part of the learning process.

The Edit Page

On the Edit page, go to Workspace > *Reset UI Layout* so we are all on the same page. Resolve's intrinsic workflow copies old-school film editing which is from left to right (past-present-future). If you're on the Edit page, you should have already loaded clips into a Bin via the Media page. (Of course, you can import media from the Edit page, too.) And true to form, you'll have clips on the far left, a Source Viewer in the middle and a Timeline Viewer on the right.

It's a natural reaction to just drag clips into the Timeline and start cutting (I often work that way), but a more disciplined approach will help keep you from turning your project into a giant pile of something. Working from an A/V script is ideal, but for a lot of personal projects, that doesn't happen often. But you should at least work from a shot list of all the media you'll need including VO and other audio tracks, stills, and stock footage—everything, or even a roughly sketched out plan on a white board. And you should color-code clips so that all the SOT are one color, all the B-roll is another color, etc.

If something is missing—say a piece of stock footage you just haven't found yet—you can put a slug in its place as a reminder (Effects Library > Generators > *Solid Color*). You'll have to change the color from black in the Inspector, and you'll have to select *Pick Screen Color* before you actually pick a screen color. For missing media, I usually pick a color (dark red) that's the similar to the Media Offline warning. If there are lots of missing pieces, it's a good idea to use the Title Generator and label them. You'll thank me later.

UI

Because of the limitations of customizing Resolve's UI, the first order of business is to kill or resize panels you are not using in order to get as big a view of the Timeline and Viewers as possible. (And do that continuously throughout the edit as you no longer need certain panels.)

At the very top left of the Edit page is a button (down arrow) that expands (or shrinks) the left side of the Timeline if Media Pool or Effects Library are selected. There's a corresponding Timeline Expander on the right as well and likewise it is grayed out unless the Inspector, Metadata, or Mixer is active.

Viewer Selector

The Viewers and Timeline are adjustable (the grabber is right above the ruler on the Timeline), as are the Track heights. I think you'll find there's enough flexibility with Resolve's UI if you take the time to set up the UI for the task at hand and change it as your needs change throughout the edit process.

Top right is the Viewer selector (if the Inspector and Metadata are off). You can choose to have the Source Viewer up or not. Typically, you'd work with both Viewers on for the rough cut, but once all the source footage has been viewed and incorporated, kill the Source Viewer and expand the Timeline Viewer. You can grab beneath a Viewer and expand or shrink it to your preference. At the bottom of the Timeline Viewer is a pulldown for Timeline View Options. On the far left (center screen) is a *Timeline View Options* pulldown you can use to display audio waveforms or the show video tracks as a filmstrip or thumbnails.

Source Viewer

The Source Viewer and the Timeline Viewer are *nearly* identical. Double click on a clip in the Bin to view it in the Source Viewer. Top left is a scaler used to adjust the size of the frame in the space available. The default is Fit. At the top left is a timecode (TC) display that represents the *length of the clip* selected in the Source Viewer. The right TC indicates the TC position of the play head within the clip.

Despite the UI's many limitations, you can turn off panels you are not using, move things out of the way, and increase the size of the Viewers. For this stage of the workflow, it's a good idea to make the Viewers larger and Ctrl+F when you need full-screen. Also, if you don't have multiple Bins, click on the "Sidebar" icon on the top left and expand the Timeline. With just the Timeline and two Viewers up, the UI, even on a Mac, is more than adequate. Ever wonder how the folks creating YouTube tutorials have such large Timelines and Viewers when your screen is mostly gray? This is how.

On the top right of the Edit page is another Sidebar controller that expands the Inspector. With Resolve 18, many functions have been added to the Inspector and you may need to see what's what. But when either the Inspector or Metadata is on, you only get the Timeline Viewer.

I&O Points

When you select in and out points, the Source Viewer's left timecode changes to the length of the segment bounded by the I&O points. If you now drag that subclip into a Bin, the name gets "subclip"

Video Only

appended to it. All subclips from the same clip have the same name, so I usually add a number to the subclip's name so I can sort them.

If you make a subclip by dragging your selection to a Bin, you'll find it difficult to increase the duration of the subclip. However, if you drag it from the Source Viewer *directly into the Timeline*, you can use the Resize cursor to lengthen it. Alt+X removes I&O points, or you can use the Selector (arrow) and just move the dots. If you drag a clip from the Source Viewer to the Timeline, you get the entire clip (as defined by I&O points, if any).

If you hover at the bottom of the Source Viewer, and the clip has audio, two icons will appear. The *Film Frame* icon lets you drag *just the video* from a clip (click-hold down-drag). The *Sound Waveform* icon does the same thing for audio. This is handy for pulling B-roll because you don't have to go into the clip and remove the audio (thus, you can make B-roll instantly out of SOT).

Three Dots

Behind the Source Viewer's three dots are a couple of useful items. Leave *Live Media Preview* enabled. With SOT, I often turn on *Show Full Clip Audio Waveform*. That lets me quickly find the start/stop points of various takes (by seeing the "claps" or where the audio breaks). As mentioned previously, under the Timeline Viewer's three dots unclick "Show All Video Frames" which will improve playback with less rendering.

Transport Controls

The Source Viewer's Transport Controls should be familiar: Go to First Frame, Play in Reverse, Stop, Play, Go to Last Frame, and Loop. The space bar is a shortcut for Play if it's stopped or Stop if it's playing. The J-K-L keys have historically been used to play reverse (J), stop (K), and play forward (L). A very handy use of these keys is to hold down the K and play forward or reverse using the J or L keys—*plays very slowly.*

To loop a clip (have a selected portion play over and over), set I&O points in the Source Viewer, click on the Loop icon (turns red), and press Alt and / (forward slash) to start. The space bar stops looping.

The next button I want to point out is *Match Frame*. This makes the Timeline play head jump to the same frame the Source Viewer's play head is on. This is good for finding exactly where in the Timeline a particular frame (or clip) is located.

Timeline Viewer

The Timeline Viewer controls are nearly identical. Here are the exceptions: Top row, there is a rainbow with stars button that lets you bypass any effects that have been added from the Edit, Color, or Fusion pages. Notice that clips in the Timeline have an "FX" icon if Open FX effects *were added via the Edit page* (but not if they were added on the Color page)—Why? IDK.

Bypass Effects

At the bottom left on the Timeline Viewer is a Transform button that gives you access to controls that will let you scale or crop the frame via a WYSIWYG wireframe. But there are similar tools in the Inspector, more of them, and they are easier to use.

The play head under the Timeline Viewer syncs with the Timeline's play head. As mentioned elsewhere, it's a good idea to set up the Edit page so that the clip under the Timeline's play head is automatically selected (Timeline > *Selection Follows Playhead*).

Timeline

When you drag your first clip to the Edit page, Resolve will create a Timeline. The configuration will be determined by the Preset you selected from the list in Project Settings > Presets or the default System Config if you didn't. As mentioned, you cannot change the *frame rate* of the Timeline once set, and although you can make other changes (some may require re-rendering the entire Timeline). But it's best to start out on the right foot.

Film School

There are two basic approaches to editing: cutting everything (or nearly everything) on one master track, or by stacking tracks vertically, one on top of the other until the sky is reached. Putting nearly everything on a master track is a nod to 35mm film production technique.

With 35mm film, there was only one video track, and everything was spliced onto that one reel of film. If we wanted a dissolve, we sent the two scenes to the lab, they printed them with a dissolve and sent us back a single strip of film (dissolving the two clips) that was spliced into the single master reel.

To edit 35mm film, the two pieces of film to be joined were put in a "hot splicer" and clamped down. One side of the film had an emulsion coating that had to be scraped off first so the cement would take. The splice was cemented and clamped into place until the heat from the splicer evaporated the solvent. The splices were invisible because they occurred in the unexposed (black) area between frames. If you are editing everything using one master track, you are basically editing like we did back in the day with 35mm film.

With 16mm film, it was a different story. The space between 16mm frames is not wide enough to hide the splice—there's not enough room—the splices would be seen when the film was projected or printed. In TV news, to get around that we used very thin clear tape and not hot splices. It was sort of invisible (not really), and typically, that film was only projected once. But for creating a 16mm negative (or positive) that could be printed we had to use cement splices and a slightly different technique.

The solution was to edit 16mm as two tracks in a checkerboard pattern using an opaque film stock called "black leader." Where there was video on the A-roll, there was black leader on the B-roll, and vice versa. When 16mm film was printed, both the A and B rolls were printed at the same time. The black leader prevented the printer's light from exposing film (and the splices) on one roll while the other roll printed. This system led to using multiple tracks for lots of things. I've edited 16mm film with as many as 6 tracks (A, B, C,

D, E, and F) to accommodate titles, animation, different film stocks, and effects. But the basic story was carried on the A and B rolls.

With 16mm film, most of the audio was synced to the A-roll. And when we wanted to make an edit, we'd splice footage onto the B-roll to cover it. BTW, that's where the term "B-roll" comes from—it's the name of the roll that has the shots you are cutting away from the A-Roll to. So when you are editing by stacking tracks, that's basically how 16mm film was edited. This is also known as "checkerboard" editing.

Which approach works best for you is a personal choice. I've worked with editors who squeeze everything onto Video 1 except titles, and others who stack tracks on top of tracks on top of tracks. Sometimes there's a method to the madness. Other times, it's just madness.

Since I come from both film and videotape, I like to work with a hybrid workflow that has both the SOT and B-roll on Video 1 and stacks titles and graphics and the occasional odd shot on separate tracks. But do it whichever way you like.

Editing—An Approach

One approach to editing is to put all the clips into the Timeline in roughly the order you intend to use them, then use the razor blade and cut out the parts you are certain you do not want. If you set I&O points back on the Media page, that shouldn't be that much. If a clip is missing, add a slug of appropriate length (Effects > Generators > *Solid Color*) as a placeholder.

Once the footage has been assembled, edit the *audio track* first. The audio track has to be *continuous* (while video is by default a *discontinuous* series of individual frames.) Also, with audio you have fewer replacement options while you can cover a multitude of sins in video (B-roll, cutaways, reaction shots, full-screen graphics, stills).

Imagine you are doing a standup and some boob walks in frame, pauses, and backs out. That's easy to fix with B-roll, a cutaway, or a full-screen graphic. You could digitally zoom in, too. Even popping in a lower thirds might make the offender less noticeable (draws the eye away).

But if the boob also makes a noise, that's much harder to deal with. You can duck it. You can edit in a piece of Room Tone. Worst case, you could ADR. Either way, your options are more limited and audio anomalies are always more noticeable.

Creating a smooth, clean, continuous track can be much harder with audio than video. So always start with the audio first. Once that's in fairly good shape, you will have much of the video editing already done, and you can work on cleaning up the main video track.

When editing the video track, just use butt splices on the first pass. Don't bother with transitions, effects, titles, or fine tuning the audio until you have a rough cut that roughly works. Otherwise, you'll find yourself spending hours tweaking clips that ultimately never get used.

Ripple, Roll, Slip, and Slide

There are tools and techniques beyond the razor blade. Mostly these tools are for working with a single master track and are not needed if you are building a video by stacking tracks. What follows is a written description of the Edit page tools. I suggest you also watch several videos that show what is inherently a visual process. While there are more than 50 YouTube videos on this subject alone, the clearest explanation (IMO) is not on YouTube but can be found at:

> PREMIUMBEAT.COM/BLOG/DYNAMIC-TRIM-TOOL-RESOLVE/

To practice with the Ripple, Roll, Slip, and Slide tools, you really don't need anything more than three clips (visually different) of about 10 seconds each.

Roll

Roll Cursor

When you mouse over the junction of two clips in the Timeline, one of two icons will appear. The icon that looks like a bridge is the Roll cursor. You can mouseover a splice, hold down the mouse button, and use the Roll cursor and move the edit point around.

In doing a Roll, you are extending one clip while shrinking another so that *the total length of the project does not change*. Roll edits are handy for dealing with B-roll because you can set exactly where the B-roll starts and stops, and if you don't like it, you can move those edit points around without affecting anything else. If the edge of a clip is green, there's more footage available but not shown. If it's red, that means you've run out of clip.

Depending on what you are trying to do, you may want to unlink the clips at this point and work with video only. That's a common practice for adding B-roll and other visuals that don't have sync sound.

Resize

Resize Cursor

The other cursor looks like a crossbow (find it on the left or right of any splice). That's the Resize cursor. With Resize, you can extend or shrink the duration of a clip. If you lengthen it, that will cause the adjacent clip to shrink. If you shorten it, a gap will appear. If there's a gap, you can extend the adjacent clip to fill it, which is essentially the same thing as a Roll, just not as fast or as precise. You can also click on the gap, and press delete, but that shifts left everything downstream of the edit and shortens the project by the length of the removed gap. You may be OK with that.

Trim Edit Mode

Trim Edit Mode

Next to the Selector cursor (red arrow) is the Trim Edit Mode (T). With Trim Edit Mode on, you also get Ripple, Roll, Slip, and Slide.

Ripple

With the Trim Edit Mode on, the Resize cursor becomes the *Ripple* tool. In the Ripple mode, changing the duration of a clip affects everything to the right of the edit. In other words, changes in the duration of the clip "ripple" down the Timeline. Change back to the Selector arrow (A) and the tool reverts to Resize.

Roll

A *Roll* edit changes the end point of one clip and the beginning point of another without affecting anything else (project length stays the same). If you Roll the edit point between clips A and B, you change where A ends and where B begins. That changes the length of A and of B as well at the point at which the edit occurs, but does not affect anything else in the Timeline.

Slide

Slide Tool

A Slide means sliding the entire clip without changing its length—Slide only changes the clip's *position* in the Timeline. To perform a Slide, grab the thumbnail of clip near the bottom. As you Slide clip B, you are simply moving where it is in the Timeline—nothing else changes. Slide only affects *where*.

Slip

Slip Tool

A *Slip*, on the other hand, is the opposite. A Slip changes *what* parts of the clip are shown within the segment used in the Timeline, without changing the clip's position or duration. Clip B may be in the correct position and of the right length, but the footage in clip B may not show the action you want. To Slip a clip, grab the "tie fighter" icon in the center of the clip and drag. Slip only affects *what* is shown, not where or when.

Example

As an example, say you have a project that's 30 seconds long and made from three clips, A, B, and C. If you want the B clip to be a little longer or shorter but you also want the project to stay the same length, you can mouse over the splice at A-B or at B-C, activate the *Roll* cursor, then drag it (to expand or shrink clip B). In doing so, depending on which side of the B clip you are on, the A or C clip will shrink or expand so that the entire project stays at 30 seconds. What you take from or add to one clip will shorten or extend the adjacent clip by the same amount so that the length of the project does not change. Roll only changes where the splice occurs.

If you *Ripple* the right side of clip B, that changes the length of clip B and moves C to take up the slack—everything to the right will move left to accommodate the change.

If you *Slide* clip B, you do not change its length or what the clip shows, just the position of clip B in the Timeline.

Finally, if you *Slip* clip B, you are not changing its length or position, just the contents of what clip B shows in the Timeline. You can use Slide to put clip B exactly where you want it, and use Slip to adjust what clip B shows.

While you should probably watch a few videos to better understand how these tools work, you really need hands-on practice with Roll, Ripple, Slip, and Slide get a feel for how to use them.

There are additional edit tools in Resolve, but most edit tools are simply time savers that let you do in one step what you could do with just the razor blade and Ripple Delete in multiple steps. That's great for repetitive tasks, but these tools can be hard to master if you don't use them regularly.

Dynamic Trim Mode

Dynamic Trim Mode works with the J-K-L keys to dynamically Ripple, Slip, or Slide clips while you play them. Not really for beginners.

Position Lock

There's also a *Position Lock* that locks a clip and won't let you (accidentally) move it within the Timeline.

Insert, Overwrite, Replace

If you already have some clips in the Timeline and drag a new clip in, the new clip will *Overwrite* what's in the Timeline (where the clip is when you release the mouse). Usually, that's in gray space and doesn't affect anything else in the Timeline.

If you need to put the clip in between two existing clips, you could *Insert* a clip and have the Timeline move things around to accommodate the inserted clip. You can do these with various "drag and drop" techniques, but I think you'll be less confused if you use the buttons.

Overwrite

Overwrite

If you drag a clip on top of existing clips, the new clip will overwrite whatever was there. If the clip also has audio, the audio track will also be overwritten. This control requires you to know what you are bringing in (lengthwise) and what you are getting rid of. Overwrite is the default. You can also Overwrite clips in the Timeline using the *Overwrite Clip* button (F10).

For a more pleasant experience, place the play head in the Timeline where you want to new clip to start. The new clip should have already been trimmed to the exact length you require. Select the clip in the Bin so that it's in the Source Viewer. Then click on the Overwrite Clip button. The new clip will start exactly where the play head is located, and it overwrites any clips in the Timeline per its length.

Replace

Replace

Replace (F11) is similar to Overwrite, but the amount of Timeline that gets overwritten is determined by the length the clip *you are overwriting* rather than the length of the new clip. This is a handy tool if you have a sequence cut and don't like a shot. You can *Replace* the bad clip in the sequence without screwing everything else up. Once replaced, you can Slip the new clip to change what the clip shows. And you can Slide the clip if you want to tweak its position.

Insert

Insert

Often, when you want to Insert a clip in the Timeline, you may want to part the waters and have everything move right to accommodate the new clip. If you put the Timeline's play head at the point where you want the new clip to start, put a new clip in the Source Viewer, and click the *Insert Clip* button (F9), the new clip will be inserted and *everything to the right of the Insert point will Ripple (move right)*. Basically, it's a shoehorn move: Space is made for the new clip and everything to the right moves out of its way.

More Icons

The magnet (Snapping) icon snaps the clips together as if the clips' edges were magnetic. It's on by default, and that's good because almost all the time you'll want it on. However, sometimes the "magnets" grab the edges of the clips and won't let you move them arbitrarily when the edges are nearly touching. If that's a problem, turn Snapping off (N). I know, should be M, right? But it's not. It's N.

The chain link icon turns *Linked Selection* on and off. Typically, you unlink clips to work on just the video or just the audio and relink them when you are done. Clips that are linked are red boxed. The audio and video tracks of a clip are linked by default. If you unlink a clip in order to work on one or the other, be careful not to get the audio out of sync (red warning TC at the bottom left of the clip). Ctrl+Z if you do. You can also right-click on linked clips and check or uncheck Linked Clips.

You can use the Ctrl and − or + to expand or shrink the Timeline (temporally). That's faster than using the slider. Putting the cursor anywhere in the Timeline and holding down Alt and using the mouse wheel is much faster and more precise.

You can rename tracks to anything you want. Video 1 could be "Master," Video 2 "B-Roll," Video 3 "Drone." You get the idea.

Each track has a Lock, Auto Track Selector, and Mute button. There's also a Disable Video Track button that kills the video (takes it to black). If the track's identifier has a red box, that track is the destination for the Overwrite, Insert, and Replace buttons. You can click on another track's ID (i.e. "V2") and change whether or not it's the intended destination. If you are dragging and dropping a clip into the Timeline, where you actually drop it determines where it ends up. And if there's no track there, Resolve will create one for you.

At the far right, there's a Volume control that just controls your speakers and not the track or Master volume and a DIM button that ducks the volume a bit—for when you need to make a quick call but still want the bosses to hear the track playing in background to prove you are still hard at it.

The Fusion Page

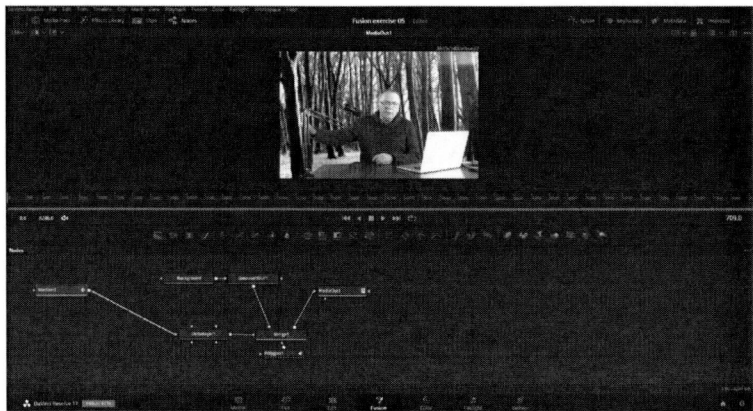

It may now live inside Resolve but, like David S. Pumpkins, *Fusion is its own thang*. BMD describes Fusion as "the world's most advanced compositing software for visual effects artists, broadcast and motion graphic designers, and 3D animators." Note that at no time does BMD ever say "for beginners and casual users"

That's because Fusion is not for casual users or those who aren't willing to spend the time and effort necessary to learn its intricacies—which are legion. Even seasoned pros often struggle with it. And like Cut page, Fusion doesn't really work like the rest of Resolve either. It uses completely different concepts and tools. And Fusion requires much more computational oomph than the rest of Resolve.

For all these reasons, I was hesitant even to include a chapter on Fusion that covered more than the Fusion page UI itself. But mostly I didn't want to include Fusion because once you see what you can do with Fusion, you'll be miffed (not my original word choice) that there's not more in the book about Fusion. Lots more. Yes, it's truly da bomb.

The real advantage is the elegance of Fusion effects. What it takes 10 steps to do via OpenFX or the Color page, you can do with a couple of Nodes in Fusion. Plug and Play. And Fusion lets you create much more sophisticated effects with less fiddling with the controls. But your computer has to have the powerful chips necessary to process Fusion effects. While there are lots of settings and operating procedures mentioned previously that help with playback and render issues in Resolve, those won't help you as much in Fusion. Here, you'll need pure horsepower.

So, as a compromise of sorts, I am going to go ahead and show you how to re-create a previous exercise using Fusion rather than OpenFX. Nonetheless, it behooves you to watch a few Fusion tutorials before striking out on your own. Blackmagic has an excellent Fusion training video and a Fusion Visual Effects book (you can download from BMD for free). CB Super has posted several first class Fusion demos and tutorial videos.

Casey Faris, Chris' Tutorials, and JayAreTV also post excellent Fusion-related tutorial videos from time to time. However, Fusion has changed quite a bit over the various releases, so look for tutorial videos starting with Resolve 16 or later. I would also point out that the section on Fusion Fundamentals in *The DaVinci Resolve 18 Reference Manual* is clearly written and (thankfully) short. In the current edition it begins around page 1076. Fusion has a lot of arcane rules and secret passages, and you can never know too much about Fusion.

Fusion Versions

There is a Fusion page in Resolve, obviously, and there's also a standalone Fusion Studio app. The standalone version of Fusion has more advanced tools and can render over an unlimited number of networked computers (such as a render farm) for near instantaneous processing. The standalone version also works with other NLEs like Avid and with multiple collaborators. If you are a game developer, a VFX artist, or are working on a 3D animated feature film for a major studio, spring for the standalone version. But the rest of us will find the Resolve version of Fusion more than adequate.

Fusion Cheats

Many people just use Fusion just to create animated titles (they are slick). However, pre-fabbed animated Fusion Titles are now available from the Effects Library on the Edit page. And certain Fusion effects like Delta Keyer are now available on the Edit page. If that's all you need, you no longer have to learn Fusion to use them.

Fusion Explained

Fusion is a *compositing and effects* system. Compositing is an old film-industry term that means taking two or more pieces of film and combining them onto one piece of film—a composite. I'm not talking about simply *combining* two video tracks (like when you put a title over a scene). Fusion can combine multiple video clips and incorporate a whole host of effects such as Chroma Key, mattes, masks, tracking, and the like, and meld the all that into a single composite clip. In that sense, it's not unlike 35mm film production. And Fusion works with Node trees instead of layers much like the Color page.

In the following exercise, you will work on a clip directly in the Timeline. This method (called "single clip composition") is a more straightforward way of approaching Fusion for the first time. Additional Fusion composition methods are discussed later.

Fusion Demo Project

Start a New Project. Import Media (the green screen clip and a background image from a previous exercise) to the Master Bin. You really only need a 10 second clip for this (shorter is better if you are going to see how your Fusion project really looks by rendering for final in Deliver).

Switch to the Edit Page, and drag the foreground (green screen) clip to the Timeline like you did at the beginning of Exercise *Due*. Put it on Video 1 and select it. Forget about the background clip for now.

Switch to the Fusion page. Turn Playback > Render Cache to *None* so you can work unimpeded—no point in constantly rendering while you are trying to set up an effect.

Go to Workspace > Reset UI Layout so we are on the same page. Grab just below the Transport Controls and pull down to make the Viewers larger or smaller as needed. (This doesn't work on the Mac.)

Notice that you already have two Fusion Nodes as defaults—Media*In*1 and Media*Out*1. Note also that Fusion Nodes have inputs (triangles) and outputs (squares) and are joined by a connecting line. The shapes and colors of the connectors have meaning–I'll get into that later.

If you hover over the bottom left of a Fusion Node, a Viewer selector will appear. The left dot is the left Viewer and white means on. Both the left and right viewers are identical—the only difference is in the signal they are being fed. (That means you can get them mixed up if you're not careful.)

Turn the MediaIn1 Node's *left* Viewer ON. Turn the MediaOut1 Node's *right* Viewer ON. This way, the input is on the left Viewer and the output is on the right Viewer.

Double click on the yellow connector between MediaIn1 and MediaOut1. It goes away. Now drag a new connector from the *square output* of MediaIn1 to the *triangle input* of MediaOut1. That was just for practice. Move the Nodes around if you want. You won't break anything.

Secret Codes

The *shapes and colors* of Fusion Node connectors have very specific functions. A Merge Node, for example, has inputs for a foreground image, a background image, and some sort of Effect Mask that tells the Merge Node what to do with the two images. If you get the connections wrong, weird things will happen or maybe nothing at all.

How many connectors a Node has and what the connectors do depends on the Node's function. You can mouse over a connection, and it'll tell you what it does. The shapes are pretty consistent but what a color means often varies with the effect. For the exercise that follows, the shape/color coding works like this:

   Green = Foreground
   Yellow = Background
   Blue = Effect Mask

White = Solid Matte
Pink = Clean Plate
Square = Output
Triangle = Input

When you connect two Nodes, Fusion will probably move the connection locations around for you to clean things up. That's normal behavior.

The Key to a Clean Green Screen

(I'm going to number the following steps so you don't get lost. Once you've done this a few times, it'll seem intuitive.)

(1) Right-click anywhere in the dark gray grid (bottom) and select Add Tool > Matte > *Delta Keyer*. Magically, a new Node called Delta Keyer will appear.

(2) Repeat for Add Tool > Composite > Merge. You'll get a Merge Node which is where the signals will get composited. At this point you'll have four Fusion Nodes.

(3) Open the Media Pool and drag the *Background image* you want to use to the Nodes grid. A new Node will appear named MediaIn2. Right-click on it (or F2) and rename it "Background."

Now that you have all the necessary Nodes for a green screen, it's time to play "connect the dots."

(4) Double click on any existing connections, and remove them so you have a clean sheet.

(5) Drag a connector from the gray output square of MediaIn1 to the yellow input triangle of the DeltaKeyer1. The output square will turn white. (Move DeltaKeyer1 around if you want, to keep things tidy.)

Drag a connector from the gray output square of the DeltaKeyer1 to the *green* input triangle of Merge1 (moving Nodes around as neccessary).

(6) Drag a connector from the gray output square of Merge1 to the *yellow* input triangle of MediaOut1. Turn the right MediaOut1 Viewer on.

(7) Now drag a connector from the *white* output square of the Background Node to the *yellow* input triangle on Merge1.

(8) Clicking on the Delta Keyer Node opens the Inspector. Find the Background Color box (ironically gray) and the eyedropper. You could use either or both to select the green screen color to be keyed, but the eyedropper is more exact.

(9) Click, **hold down**, and drag the eyedropper to the green screen.

You should now have a green screen composite. If you only see the background, your wires are crossed. In that case, right-click on Merge1 and select Swap Inputs (Ctrl+T).

Strangely, the Background image's *resolution* can sometimes affect the size of the foreground image, and the foreground image may need to be resized. If that's the case, select Merge1. On the Inspector, adjust *Size* until the image fits the screen (using the slide control or by grabbing the green box in the Viewer). *Reset* (circle/arrow) if you need to adjust.

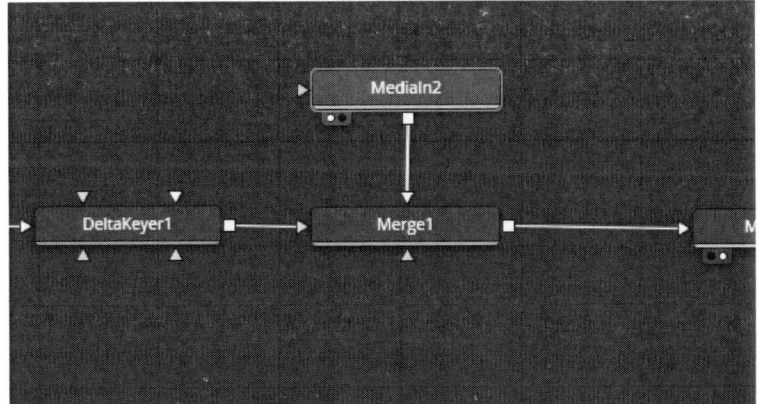

Fusion Exercise Nodes

Adjustments

Usually, the above steps are all you need to do to create a clean green screen. If you need to make fine adjustments, select Delta Keyer. Selecting the *Matte* brings up controls similar to those on the Color page. You'll find that Clean Foreground and Clean Background can work much better than Clean White and Clean Black on the Color page because you can see exactly what you are doing if the View Mode is set to *Final Result*. Adjust the Matte controls as necessary. For removing spill, select *Fringe*. Spill Method defaults to None, so set it to something and adjust *Spill Suppression*. And you can always Reset and reapply the eyedropper.

Rendering/Caching Fusion

Turn on Render Cache > *Smart* and let it do its thing. Play the composition back from the Edit page.

With Resolve 18 and a modestly endowed computer, you shouldn't have problems playing Fusion compositions (unless they are very complex). Fusion has its own rendering system and uses a different control scheme, but it is controlled by the Timeline Render Cache (Smart/User). The Render Cache Fusion Output used to be called Render Cache Clip Source and supposedly renders everything up to and including Fusion (which means it doesn't render effects done on the Color page).

If you do experience playback issues and your footage is H.264/265, using Proxy or Optimized Media (to DNxHR or ProRes) before building the key may be of help. You could also try right-clicking in the Fusion Transport Control area and turning off High Quality and Motion Blur.

However, with Resolve's *Render in Place* (right-click on a clip to reveal), you could force Resolve to render and cache the problem segment. That could take a while, but the result will play smoothly. If you need to make changes, you can always Decompose to Original and redo.

For me, jumping to the Edit page and playing a Fusion clip from the Timeline works better than trying to play the same clip on the

Fusion page. This behavior is probably due to the difference in how the render systems work. Earlier I mentioned that Render Cache does just enough processing so that the Timeline will play smoothly. Render Cache Fusion Output (and Render Cache Color Output) takes things further–all the way, really. The "output" render caches process clips to the point where they are ready for output. That takes longer than processing just for playback.

Garbage Matte

Creating a Garbage Matte in Fusion is similar to Color. But it's different, too. On the Fusion page, go to single Viewer Mode (little box on the Viewer's top right). If you need room, adjust the Viewer so that the image is surrounded by black (click on the Viewer then Ctrl + mouse wheel to zoom).

Right-click in the Nodes grid and select Add Tool > Mask > *Polygon*.

On the viewer, click to set the corners of your Garbage Matte (which will mask out the *non*-green screen areas). I think you'll find it easier if you create a rough wireframe, setting lots of adjustment points (clicking as you go), and then adjust the resulting wireframe to fit. You can add more adjustment points by clicking on the wireframe. Make sure the wireframe closes completely (makes a closed box).

Connect the Polygon1 output square to the blue connector triangle on the Merge1. Click *Invert* on the Inspector > Controls, and adjust the wireframe as necessary.

Effects

Select the Background Node. Press shift-space. You'll get a menu (useful for other things). Type in "blur" and select Gaussian and *Add*. Make sure the right Viewer is on for MediaOut1. In the Inspector, adjust the Strength to whatever looks good to you.

You can add an effect to any Fusion Node by selecting the Node, clicking shift-space to bring up the search window, and searching for an effect that suits you. And you can tell at a glance what Fusion

Nodes are doing, and adjust on the fly. (You can also preview effects live on the Edit page).

On the Timeline, clips that have Fusion effects also have the *three stars icon* at the lower right.

There are hundreds of effects tools in Fusion. You can hunt for them by name like you did earlier, or you can use the row of icons (mid screen) that will get you to the most commonly used Fusion effects. Some are arcane and are known only to a few folks at DreamWorks or ILM, but there are some that the rest of us may find useful. Mouse over the center control ribbon and locate these: Background, Text, Paint, Color, Blur, Merge, Matte Control, Resize, Transform, Rectangle, Ellipse, and Polygon. Clicking the icons is the same as wading through the menus.

When you've finished, go back to the Edit page, let it render, and make sure the Timeline plays perfectly. Then do a File > Quick Export (H.264) and see what it really looks like.

If after this exposure you want to really get more into Fusion, watch Casey Faris' "Fusion: The Ultimate Beginners Guide." IMO it's the best intro to Fusion on YouTube. CB Super has a flight of Fusion tutorial videos for beginners that are also excellent.

Fusion Composition

When you select a clip on the Edit page and jump to Fusion and apply effects, you are essentially creating a Fusion composition linked back to the clip on the Edit page.

Instead, you could generate a blank Fusion composition (Effects Library > Toolbox > Effects > *Fusion Composition*) in the Timeline and later add media and effects to it. The default length of the generator is 5 seconds but you can expand it. Most of the time, I find it simpler to apply Fusion effects directly to a clip, but there will be times where you need to create a placeholder Fusion Composition in the Timeline first and add media and effects later. In fact, try it now. Add a Fusion Composition clip and repeat the exact green screen exercise you just did, starting with an empty Fusion Composition clip. (Hint: Grab the green screen shot first and it will appear as MediaIn1. The rest of the process is the same.)

Fusion Compositions vs. Compound Clips vs. Fusion Clips

In the previous exercise, you created a Fusion composition linked to a single clip. And, if you were diligent, you repeated that using a Fusion Composition clip.

As you recall, on the Edit page, you can merge multiple clips into a single Compound clip and work on it as one thing. So in theory, you could use a Compound clip on the Fusion page. But there are issues with doing it that way. The Compound clip appears to Fusion as a single MediaIn Node, and while that may be OK, it also means you can't perform Fusion operations on the constituent clips unless you first decompose the Compound clip in the Timeline.

What you need is a process that lets you apply Fusion effects to a single compound clip and also the individual constituent clips. The Fusion version of the Edit page's Compound clip is called a *Fusion clip*. A Fusion clip is simply a Compound clip designed to work with Fusion, and it allows you to also perform Fusion operations on the individual constituent clips. (But ironically, you can't use Fusion titles (or generators) in a Fusion clip.)

Obviously, whether you need to work with clips as Fusion compositions, Compound clips, or Fusion clips depends on many factors. And knowing the answer in advance takes lots of practice working with Fusion. For most casual users, selecting a clip in the Edit Timeline and switching to the Fusion page (and in the process creating a Fusion composition) will often suffice.

The Color Page

So much of what happens on the Color page was covered earlier in the exercises, that I'm going to focus mostly on things that might have been missed plus some of the controls that didn't even get honorable mention.

Bypass Effects

Above the Timeline Viewer is the Bypass button (in color) that turns Color, and Fusion effects completely off (or back on) except those effects applied from the Edit page. The default is on. Next to that is the Enhanced Viewer (expander). Try it out, but leave it off for now.

Note that if you need to remove an OpenFX effect that was applied on the Edit page, right click on the clip, select *Remove Attributes*, then select *Plug-ins*.

Highlight

On the far left of the Timeline Viewer is a "magic wand" Highlight button that emphasizes the particular effect you are working on. When enabled, three additional controls appear (on the far right...). These controls are Highlight, B&W, and A/B (difference). You used B&W when you were cleaning up the Chroma Key in the exercises. It's important to use B&W when adjusting the Matte Finesse controls. Too many tutorials give rules of thumb for these settings (and no underlying explanation). By using Highlight and B&W, *you can see what the controls are actually doing.* The A/B mode is very useful for color grading. Nathan Carter's YouTube video on "Ten Resolve Color Page Features" explains how to do this.

On the Color page, you can view thumbnails. But there is also a mini Timeline that is hard to use and rarely needed. Fortunately, you can turn that lesser Timeline on and off (click *Timeline* at the top right). The clips in thumbnail view are in Timeline order anyway, and you'll usually be applying effects at the clip level anyhow. Ditto for *Clips*. You don't have much say in how the UI presents information, so manually turn off what you don't need to see when you don't need to see it.

Color Wheels

These controls, as you should know by now, adjust the Chroma and Luma in three broad bands Lift (shadows/blacks), Gamma (mid tones), and Gain (highlights/whites). *Offset* controls the whole shebang (Lift, Gamma, and Gain) in one control. *Color Wheels* lets you control RGB and Luma with the Wheels. You can change that tool to Bars or Log. Bars is more visually oriented while Log is more subtle.

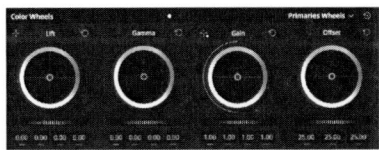

Color Wheels

The Color Wheels are surrounded by various finesse controls. This is what they do:

Temp

Adjust the overall Hue of a clip based on color temperature (Kelvin). Thus, you can "warm" things up a bit.

Tint

A Hue control that should be familiar to you if your family had a CRT-type color TV.

Contrast

Contras lowers Lift and raises Gain without affecting Gamma (that much). Basically, it makes everything more "contrasty."

Pivot

Pivot sets midpoint of the Contrast range. Use Pivot in conjunction with Contrast to hone in on just the right blend of light and dark.

Mid/Detail

Increases sharpness in mid tones.

Col Boost

Increases the Saturation only in low Saturation areas. More useful than Sat because you can increase Saturation just in the areas that really need it without turning the picture into Technicolor (unless that's your goal).

Shadow

Used for getting detail from the Shadow areas.

Hi/Light

Used with HDR media to bring down blown-out highlights without affecting Gamma too much.

Sat

Color Saturation control (many colorists increase Saturation via Col Boost instead).

Hue
Yet another way to change the tint.

L. Mix
L. Mix (defaults to 100) automatically adjusts Luma based on changes you are making to Chroma. So if you adjust blue, Resolve will automatically adjust red and green such that the overall Luma remains constant. Sometimes, that's OK. If not, lower the value all the way to zero and totally separate Chroma adjustments from Luma. Or somewhere in between.

Color Controls
You'll have to play around with these color controls until you get a feel for what they do. It should be obvious that anything the color correction tools can do could be probably be done simply by adjusting the Color Wheels (or Bars). But individual color adjustments usually change Luma at the same time (unless you turn L. Mix down) or affect some other related control.

These hybrid controls attempt to combine several actions in one control to achieve a result without having to go back and forth using the Lift, Gamma, and Gain controls. However, you still may want to go back and tweak an individual control even after using these tools. That's usually done in the Final Node to perfectly match the scenes before and after. I should mention that the control options are what they are to coincide with various physical control knobs, buttons, and trackballs on BMD's control panels (hardware company, remember?).

Caching
A clip can be manually rendered by "flagging" a clip (by right-clicking on the clip and selecting *Render Cache Color Output*). This enables caching transitions, Edit page effects, Fusion effects, and anything on the Color page whenever Render Cache runs (Smart or User). Basically, it's caching the output of the Color page and all previous pages whether anything has been added via the Color page or not.

Effects, grading and other changes made via the Color page are applied via Nodes. When you render a clip, you are basically rendering and caching all of the Nodes. But you can also render and cache individual Nodes.

Right-click on a Node, select *Node Cache* and one of the following:

Node Cache Auto–This setting renders the Node and any Nodes upstream of it when Smart Mode renders.

On–Renders the Node whenever Smart or User Mode renders. This is the setting to force a manual render of a Node.

Off–Node is not rendered. Typically this setting is used to disable rendering of Nodes that really don't need to be rendered.

For example, some computationally intensive effects (Video Noise Reduction) could be applied at any point in the chain of Nodes. If these computationally intensive effects are added first and cached, they won't have to be re-rendered when other Nodes are added later. This doesn't work if there are Nodes upstream that still need to be rendered or if the input to the Node changes. And caching individual Nodes based on complexity and position is somewhat like playing chess or Go. But it can be handy to add computationally complex effects first and render/cache them. When you add Nodes downstream, the previously cached Nodes won't have to be rendered/cached again and again.

Main Color Page Tool Box

Color Tools

Curves

Curves uses a graphical UI to do by shape what you could do using the color control knobs.

Under *Custom*, you can change one value based on another value, such as Hue Vs Sat or Lum Vs Sat. The latter one is often used to adjust the Saturation based Luminance. For example, you might want to saturate Chroma in the mid-tones but not in the highlights or shadows. Or you can change the Hue based on its Luma value.

Curves are a set of tools that you'll need to watch several YouTube tutorials and practice with them to understand how to use them. Many colorists spend a lot of time with these controls because they are both elegant and powerful. If you want to make your camera look much better than its price tag would otherwise suggest, Curves would be the place to go. It will, however, take some study and much practice, but the results can be stunning.

Color Warper

Color Warper is a spider-web-based color control. It was new with Resolve 17. The Resolve 17 release video went on and on about it, so it must be very special. However, I'm using 18, and I still haven't found that I ever needed Color Warper. The traditional color grading controls native to Resolve seem just fine and are less complicated.

Qualifier

You should already be quite familiar with this one from the exercises—it's the eyedropper you used earlier. When you selected a pixel on the green screen, a range for Hue, Saturation, and Luma also popped up. You could have moved these sliders around to expand the Hue and catch any variations in green, or expand or contract the Sat and Lum and adjust for lighting differences. Experiment. If you are having trouble pulling a clean green screen, try adjusting Hue, Sat, and Lum (expanding/contracting the range and shifting it as needed).

Window

Used to be known as Power Window. I will again mention that the still-unlabeled mode buttons are Invert and Mask. Usually used with an alpha channel.

Tracker

I would again remind you to track only those dimensions that you actually need to track (usually just Pan, Tilt, and perhaps Zoom). You do not have to start Tracker at the first frame of the clip and will probably achieve better results if you start in the middle and track forward and reverse. If Tracker unlocks, reposition the Window and start the track again. You can set a Window just on a prominent feature to give a good, clean track, then adjust the Window to encompass the entire object.

Magic Mask
Intuitive Object Mask lives here.

Others
There are a few more one-button controls on the Color page, but I think these are best accessed from the Inspector.

Resolve FX
There are oodles of effects in Resolve FX and more than 100 pages in the Reference Manual about how to use them. Most of the controls and adjustments are obvious. Some are arcane. Some don't work at all like the Reference Manual says. Take a (very) short clip and practice with various effects to see what they really do. Here's a list of a few standouts:

Resolve FX Blur—various blurs. Gaussian is the standard.

Resolve FX Refine—Beauty and Face Refinement.

Resolve FX Revival—Object Removal and Patch Replacer are the most useful FX IMO.

Resolve FX Sharpen—Sharpen and Sharpen Edges (useful if used discretely).

Resolve FX Stylize—Vignette if you're going old school.

Resolve FX Texture—Fake film look that actually looks fake.

Resolve FX Transform—Camera Shake (I don't need it because I can do that just fine in-camera).

Copying Grades/Nodes

Copying the grade from one Node to another is easy. Simply select the Node you want to copy from and Ctrl+C. Select another clip in the Timeline and Ctrl+V.

Or select several clips, then hover over the clip with the Node you want to copy from and press the center mouse button.

Lightbox (top right) gives you a better (larger) view of the clips than just thumbnails.

Loop works somewhat differently on the Color page. Here you can set I&O points in the Timeline and loop a sequence spanning multiple clips. Simply set I&O points, click Loop, and press Play (none of that Alt backslash stuff).

Fairlight

Fairlight is a complete Digital Audio Workstation (DAW) that has lots of tools to help you create great audio tracks. Often that means fixing problems with audio tracks—ambient noise, extraneous sounds, peaks that clip and splatter. But it is much easier to produce a great track if the original recording doesn't have those problems to begin with. That lets you use Fairlight to *enhance* your audio tracks and do amazing things with sound rather than just using it for repairs.

Like the chapter on Fusion, I'm covering Fairlight because it's there, but if you are new to Resolve, I'd advise sticking to applying audio effects on the Edit page. Since you can now apply many Fairlight effects from the Edit page or from Fairlight directly to a clip, as well as to a track via the Mixer, it's easy to unintentionally double up on effects and achieve weird results. For these reasons, either apply audio effects on the Edit page and don't use Fairlight at all or vice versa.

Fairlight Page

Go to Fairlight, reset the UI and follow along. At the top is the Media Pool and also the Effects Library. You can apply effects to a clip like you did on the Edit page.

The Sound Library has nothing in it yet. But it will take you to BMD's website where you can download a free Sound Effects Library. Do that. You'll get 1.5 gigs of free sound effects—worth every penny if you need the sound of an "aluminum can being crushed" or "blood oozing." I don't know that you will ever need "Zippo lighter opened" either, but it's there if you do. Of course, you won't see any of these sound effects until you do a *.* search. (From the sarcasm you should get that IMO it's not worth the disc space...)

Double-clicking on a sound effect clip selects and plays it. The audition button puts the sound effect in the Timeline so you can listen in context. To make that work, right-click in the dark gray area below the audio Tracks and *Add Track* (stereo). Select the new Track. Find an effect you want to audition, and click Audition. That puts a sound clip in the Track right where the play head is. You can play it using the main Transport controls. It's a clip, so you can move it or edit it if you want.

Main Features

The Mixer panel on Fairlight (bottom right) is nicely done (and should be very familiar to ProTools' users). As you add additional tracks, grab the left side of the Mixer to expand it.

Open the Effects Library, and you have a host of choices. You can apply any of these audio effects to a clip. Effects applied directly to a clip do not appear in the Mixer's Effects panel, so it's better (IMO) to click on the Effects button on the Mixer and get your kicks that way. If you apply effects directly to a clip, you'll have to turn on Effects in the Inspector to adjust various parameters after the initial set up.

But applying effects to a track applies the effect to the entire track (not just a clip), and that may not be what you want. On the other hand, you can have virtually an unlimited number of tracks, so you can let every sound source have its own track.

The next button (we're still at the top) is ADR which stands for Automatic Dialog Replacement. Say you are recording a standup and a plane flies overhead. You could use ADR to watch the video and re-record the audio in perfect sync, replacing the location audio entirely. ADR is done on every film you have ever seen (at least any made since the 60s).

Next are meters capable of handling 40 Tracks (at least). I've never used more than six on a Resolve project (moving to ProTools in the rare cases where I've needed more). But they're there if you ever have the urge. In fact, Fairlight will now let you have up to 2,000 tracks. (Sorry, but you are limited to only 1,000 tracks if you don't have a Blackmagic Fairlight control panel.) Sarcasm aside, I usually kill the Meters and use just the Mixer.

Bus1 and Control Room show you the volume of the Master Tracks (but so does the Mixer).

The tiny little Viewer at the top right can be expanded via the Floating Window icon at the bottom of the Viewer. You can move it around too. Like Ivory soap, it floats. The Dock icon (top right of Viewer) puts it back in its place.

If you click on Effects library, you'll find an array of Fairlight effects. However, these same effects are present on the Edit page (Effects Library > Audio FX > Fairlight FX), and the Edit page interface is easier to use. Hint. Hint.

Recording Audio

You can record audio directly in Resolve. First, you've got to get your computer to recognize your mic is attached. How that's done depends on your computer's OS. You can use a USB mic or a standard analog mic, but you can't use a condenser mic that requires a 48V phantom power supply unless you have a separate 48V phantom power supply since computers don't offer that option.

If allowed, set your computer's sample rate and bit depth. For most narration, 48,000 samples per second and 16 bits is more than sufficient.

Now add an *Audio Track* (stereo or mono) and expand the Mixer so you can see it. On the Mixer, click on the Input bus for that Track (it will initially read "No Input"). You'll get a patch panel where you can direct your input (mic) to the Track. For the Source, highlight the microphone (left click). The Destination defaults to the Track you selected. At the bottom of the panel, click *Patch*. The input and outputs will be highlighted to show they are connected.

Plug in headphones to avoid feedback. You can Mute ("M") all of the other Tracks while you record or use headphones and listen to one Track while you record another. On the selected Track, click the "R" (arms the Track for record *and turns on the mic*).

Listen to the room and fix any ambient noise problems. Place the mic appropriately and practice your narration while setting the level on the Mixer. Finally, hit the round *Record* button on the Transport Controls (and always remember to record a bit of Room Tone first). The Timeline will be playing during recording, which allows you to watch and comment on the video as it plays (mute the other tracks if you don't want that). The normal Transport control buttons are used to Stop and Play.

Loudness

I'm sure that at some point you've heard TV commercials that were "too loud." For obvious reasons, advertisers want their commercials to stand out. To quote philosopher Ronnie Van Zant, advertisers want TV stations to "turn it up." And TV stations did. A lot. Too much, apparently.

Eventually, the "loudness wars" got out of hand. Viewers started complaining to their elected officials, and in 2012, Congress passed the CALM Act. Basically it says that broadcast and cable networks cannot play commercials that are louder than the program itself. Of course, no member of Congress knows anything about audio engineering (or much else either), so the FCC has had a whale of a time trying to enforce the CALM Act because determining what "too loud" means is a real chin scratcher.

One reason is the difference between peak sound power and average sound power. It's the average that determines if we perceive a peak sound as too loud or not. So limiting peaks doesn't prevent a commercial from sounding louder than the program if the program's average audio level is a lot lower than the commercial's average level.

The CALM act doesn't apply to the Internet, but its principles are worth noting. In making a video, you generally want the audio to be loud without ever causing distortion anywhere in the audio chain. Loud is good because a viewer can always turn it down—all the way down to mute if necessary. They can't always turn it up. If a program has enough loudness, it usually has a lot of dynamic range too. An audio track that has an intrinsic "loudness" just sounds clearer and better, even at lower listening levels.

Despite Van Zandt's admonition, you can't just turn it up. The peaks will then be too hot and will clip or distort. However, by using audio effects like compression and limiting, you can increase the *average volume* of your track, achieve maximum dynamic range, and still produce a signal that's acceptable everywhere with no clipping or splattering.

Recording Dialog/Narration

Audio is often the red-headed stepchild of video production. It's true that television is a *visual medium*, but sound quality is very important, particularly on social media channels. As I'm writing this, I just finished watching a YouTube video of one of my favorite historians. The lighting was good. The camera was more than adequate, but they recorded the program using the built-in camera mic. You can hear the reflections from the table and walls almost as much as the presenter himself. And no one rode gain.

A splatter or two here and there is one thing, but the audio was hitting 0 dB on nearly every other syllable. I checked the comments, and sure enough, there were scads of complaints about the *quality of the audio*. Right after that, somebody sent me a link to a YouTube video Vanity Fair did with Anna Kendrick. The lighting was exquisite, and her makeup flawless, but the audio–from the built-in camera mic–sounds like it was recorded in a washtub. A camera-mounted shotgun or boom would have been better. Audiences will put up with video shot under less-than-optimal conditions but are far less tolerant of bad audio. Since the audio chain starts at the microphone, let's begin there as well.

Voice Overs (VO)

If the vocal track is 100% VO, then a high quality-lapel or desk mic is preferred over a shotgun or any other camera-mounted mic. You may only own a shotgun, but if so, for this type of work, get it off the camera and closer to the talent.

The mic's main pickup pattern (typically cardioid) should be aimed at the upper chest to pick up a richer sound (chest resonance). Pop shields are really as much for protecting the microphone's diaphragm from spittle than actually reducing plosives. I still use them anyway, but a little distance or even off-axis is often a better solution if the talent habitually pops their "P"s.

The back-side of a cardioid mic is "dead," so the rear end should point toward any noise source you can't completely kill during recording. Listen with headphones and adjust the mic's placement to minimize noise pickup.

Foam windshields are also of little use except for protecting the mic. If it's windy, a "dead cat" windscreen does help somewhat. There are a lot of DIY instructions for making these from readily available materials (an actual dead cat is no longer needed, sadly). A blimp is much more effective (and 10X the price).

Adding echo is easy. Removing it is hard. And uploading a presentation with lots of "slapback" can ruin the entire effort. It's best to stay as far away from hard surfaces as possible. But if you can't, use sound-absorbent materials to minimize picking up room reflections. Locations with highly reflective surfaces can be tamed by placing moving blankets on floors or walls. Remnant carpet samples are cheap (or free) and do a great job of soaking up extraneous sounds. Sound Blankets are denser and absorb better but cost more.

Your track will consist of the talent *and the room*—with all its ambient noise sources and reflections–so before recording audio, pop on a pair of headphones and listen to the *room*. Do what you can to minimize problems before you roll sound, and you won't need to spend time fixing the audio in post and can use Resolve's prodigious bag of tricks to make a good track even better.

Ransom Notes

It has long been a common practice in TV news to record the reporter with a hand-held mic on location for the "stand up" intro and outro, then have the same reporter record a VO later in the studio using a desk or boom mic. Many YouTube videos are also shot that way.

When using two recording setups at two locations, the difference in audio characteristics can be quite noticeable to the point of being off-putting when checkerboard edited. Some have likened it to the audio equivalent of a ransom note (they kind with the letters cut from magazines). A better practice is to record the VO with the *same mic* and at the *same location* where the SOT was shot—basically identical conditions. Otherwise, the differences in the two recordings will be audible, and the differences will be even more pronounced when you alternate back and forth between the two recordings. Yet I often hear YouTube videos—even those about television production—done that way. Ouch.

Yet when recording SOT, it's a good practice to use two mics—primary and backup—but usually, only the primary usually gets used. The second mic is for safety. And here, I don't just mean if the primary mic goes dead. The safety track can also be recorded at a lower level so snippets can be spliced in if the primary track clips. And, of course, you also have a second mic if the first one actually does go dead.

For on-camera talent, the primary should be a wireless lapel, if available, or at least a wired lapel. Shotgun mics are sometimes necessary, but even expensive ones often don't compare to a properly worn lapel. And wired lapel mics are cheap. The lapel mic should be placed so that clothing won't rustle against it and random hand gestures won't touch it. The backup/safety could be a desk mic, a camera-mounted shotgun, or a boom. Yes, this results in a two-mic setup, but it's usually not objectionable when the safety track is only used to replace a splatter or bump here and there.

A practical alternative to using two mics is to use one lapel mic and record the same audio on two separate tracks with one track recorded at reduced volume (in case the talent gets too excited). If the primary track clips, you can cut in a piece of the safety track.

Professional narrators can give you practice levels that will closely match their actual performance, but most amateurs and even experienced public speakers often get wound up during recording and frequently get a lot louder than their "Testing 1, 2, 3" would suggest. It's much better to have the talent practice the actual performance, set levels based on that, and still record a second mono track at a reduced level just in case.

The camera recording level depends on the particular reference the camera uses. Generally speaking, set the volume high-ish but low enough that the peaks *never* clip or splatter. In practice, that means not over 75% of full scale (peaks, not average).

The camera operator should wear headphones and monitor the audio as closely as they monitor the video. Of course, a second crew member for that role, particularly one who knows what to listen for (clipping, splatter, planes, cars, motors, people, dogs, mobile phones), is preferred. Just last week, I got back and discovered my audio had a lot of snap, crackle, and pop (bad connector) because I don't always do this myself. So do as I say, not as I do, as they say.

Usually, it's not easy to adjust audio on a camcorder or DSLR during recording, so a separate mic mixer can be a real help, particularly if someone is available to ride gain. And a separate recorder is a really good idea (backed up by the camera for safety).

If possible, turn off HVAC systems, fans, mobile phones, and other noise sources during recording to help reduce ambient noise. As mentioned previously, recording several seconds of Room Tone can help later with editing. Use the mic's low cut/high-pass filter (if available) to reduce low-frequency rumble even further.

If you are recording a Zoom presentation, it's a good idea to have the distant talent record their audio using a real mic and a digital recorder and FedEx the resulting SD card to you. You can sync that audio in the Timeline and have a much better sound than any mobile phone is ever going to give. This is how NPR achieves such pristine audio during phone interviews: they ship digital recorders in advance to phone-interview subjects and edit that audio in later.

If the speaker is using the room's PA, make sure it's turned down enough so that you don't get feedback or "ringing." (Ringing is the noise a PA makes right before the feedback starts.) These are audience killers, so avoid them at all costs. You may be tempted to take your audio feed directly from the PA amp. Sometimes that works, other times, it doesn't. The PA mixer may have to adjust levels to accommodate other speakers or because of feedback, and that will affect your levels as well. I'll take a PA feed if it's the POTUS, but everyone else gets a mic.

Fairlight Exercise

What does all of the above advice have to do with Fairlight? *Everything*. If you are not taking pains with your audio recording, there's no point in using Fairlight.

So assuming you have recorded a clean track, put the clip (narration or a stand up) in the Timeline and open Fairlight. The first thing to do is to adjust the Track volume such that the peaks are hitting around -10 dB or so. Turn off any audio processing by your outboard mixer (if used) or amplifier. You want a flat signal chain.

EQ

Double click on the Mixer's EQ button for the Track you just recorded. (This time, the EQ button does something.)

One of the main uses of EQ is removing unwanted sounds by lowering the volume in specific parts of the spectrum (i.e., by frequency). The stock EQ has six bands. Click on Band 1 and turn it on. Band 1 defaults to a low cut/high-pass filter. Play the track and listen to the effect of changing the frequency response of the filter. A high pass/

low cut of 80-100 Hz is fairly standard. It won't completely kill all the rumble, but it helps.

Turn on Band 6. The default low pass/high cut frequency 12.9K is probably right for narration tracks (although you may want to move down even more if the voice is too sibilant. However, if you drop it to 6K or lower, the track will start to sound like it was recorded over a phone. For music, you'd want the high-frequency filter set to 15k or higher.

The next step is tricky. What you are going to do is find a "ringing" or muddy sound quality *in a specific frequency* on the vocal track and reduce its volume. In some circles, this is called a "Wolf note." Technically, in a stringed instrument, the Wolf note is the frequency where the cavity of the instrument resonates in an off way (sounds very funky—sometimes described as "muddy"—sort of like a kazoo). Like many musical instruments, the human head also has resonant cavities (sinus passages, the mouth itself). Almost everyone has a resonance in their voice at some frequency that would sound better if it was played down. This technique can really clean up and improve narration tracks.

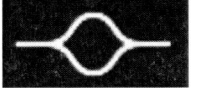
Bell Filter

The middle bands (2-5) have several types of filters: High Shelf, Bell, Notch, and Low Shelf.

Turn Bands 3-5 off for now. In Band 2, select the Bell filter. You'll get two new controls: Gain and Q. Q is basically the width of the filter. For now, we want high Q (narrow filter) so set Q to 10.3 (max).

Grab the 2 on the graphical display and drag it to about the 250 Hz mark at about +10. Now play the track and listen carefully while you slide the filter left and right. At some point, the audio will sound like a kazoo or like it was recorded in a sea shell—you'll know it when you hear it.

When you find the exact location of this muddy resonance "Wolf note," lower the Q (broadening the filter a bit). Then, lower the Gain on the filter to -5 or even -10. If these steps are unclear, there are dozens of YouTube videos that will let you hear what's going on and show you how this works. One that comes to mind is "How to EQ Like a Pro" with Alex Knickerbocker. Another good one is "How to Get Rid of Harsh Frequencies in Vocals" by Ken Lewis. Removing the muddy, nasally, funky Wolf notes in a vocal improves clarity and punch. This is a subtle effect, but it really improves the quality of vocal tracks.

You can also use this technique to reduce the level of any noise that happens to have a particular frequency (fluorescent lights, transformers, motors, fans). Simply set one of the filter bands as described above and slide it around until you find the right frequency (raising the volume makes it fairly easy to locate). Then lower the volume and adjust the filter's Q (width) so that the noise is less of a problem.

Noise Reduction

Hitting the track with some EQ can help remove some "noise" components. The next step is Digital Noise Reduction. Select the audio track. On the Mixer, click on Effects + (plus) and select Restoration > *Noise Reduction*. Now click on the *Auto Speech Mode*. Play the track and listen with Noise Reduction on and off.

Of course, there will be artifacts. While the default settings work pretty well, if you want, you can fiddle with the Threshold, Attack, Sensitivity, and Ratio as well as Smoothing and Dry/Wet to try and improve noise reduction or reduce the artifacts even more.

If you recorded Room Tone at the beginning of a clip, you can put the play head at the beginning of the Room Tone, select *Manual* on the Noise Reduction plug-in, press *Learn,* and play the clip. Before you get to any vocals, press the Learn button again to turn it off. Now Noise Reduction knows exactly what the noise in your track sounds like, so it can remove more of the noise with fewer artifacts. The takeaway is always capture Room Tone if you are in a noisy location and can't fix the noise problem on set.

Controlling Volume

You should adjust the mixer's faders so that your narration *never* exceeds -5 dB (certainly never hitting 0dB). I usually aim for -10 dB peaks. This is one case where more is better—headroom, that is.

However, when you reduce the overall volume of a recording in order to get rid of the peaks, you are reducing the volume of *everything*—so the whole track will not be as loud. Depending on the dynamic range (variation in volume), the track may sound weak.

What you really want to do is *raise* the average volume so the track will have clarity and "presence" and will match the levels of other videos riding on the same platform (YouTube) without ever hitting peaks that cause clipping and distortion. Digital is sometimes worse than analog in that respect. With analog, the system often had some intrinsic headroom to spare, and when that was exceeded, analog tended to handle extremes gracefully. But with digital, once you exceed 0 dB, the software and hardware really doesn't know what to do with the excess, and the results are far worse than overshooting on an analog system.

Of course, you can't achieve both goals using just the volume controls on the mixer. You need more sophisticated tools for that. And you also need your monitors (speakers) set to a comfortably high listening level so you can hear what you are doing. One way to achieve both goals (high average levels and tamed peaks) is to use a Compressor/Limiter.

Compression

Compression used to be a black art. Geoff Emerick built a career around his ability to fine tune the Fairchild 660, one of the first tube-based audio compressors (the 670 is the stereo version). Among his many clients: the Beatles.

It's safe to say that the Compressor/Limiter plug-ins in virtually every DAW are trying to model how the Fairchild 660 sounded back in the day. Fairlight is no exception.

> **PLOT SPOILER**: What follows is a rather involved description of dynamic audio effects. If you want, you can skip the laborious details, install a free VST plugin (LoudMax) and be done with it (see below).

Myths and Misunderstandings

A lot of people misunderstand audio compression. Some think that compression adds loudness and "punch." It doesn't. And the second I say that to anybody interested in the history of music production here in Nashville, I get the evil eye. "Of course it does," they say. "What about Brian Wilson or Geoff Emerick?" says one. "What about Joe Meek?" says another.

My point is not that Brian Wilson, Geoff Emerick, or even Joe Meek (you'll have to Google him) didn't achieve punch, a certain dynamic, a "sound," if you will using compression. They did. But the Compressor *allowed them to add the punch* by raising the overall volume of a track (on individual instruments) *without hitting peaks* that would otherwise clip and distort. Any idiot can turn up the volume. It takes a real artist to do that in ways that sound good.

By itself, a Compressor actually *lowers* the volume. After all, it's a *compressor*—it compresses the loudest parts *down,* not up. Depending on the control settings, it does nothing to the softer parts. And once the Compressor has done its job and the peaks are under control, the average volume can then be *increased* without clipping or splattering.

This section will teach you what a Compressor actually does, what it does not do, and how to adjust it properly. In order to make this subject easier to understand, instead of working with decibels, which are in negative logarithmic numbers, let's instead think about audio using a linear audio scale from 0 to 100 where 0 can't be heard and anything over 100 causes clipping, splattering, or some other unacceptable distortion.

Now imagine an audio track—a talking head vocal or VO track with an average loudness of, say, 50 while the peaks are exactly 100 and no more. Fine (probably).

But what if the average loudness was only 30, and the peaks were still 100? You can't turn up the volume and raise the average because the peaks are already at 100. So you use a Compressor to reduce the peaks *so that you can turn the volume up* (which is usually done in the Compressor plugin itself and not in the mixer, BTW). If you do not turn the volume up when you add compression, the peaks will be lower, but your average volume will have *dropped* slightly. And that's the opposite of what you want to achieve. You want to compress the peaks so you can turn the average volume up without clipping or splattering.

To activate Fairlight's Compressor/Limiter, double click on the Dynamics panel in the Mixer, and you'll get 21 controls to play with. Nice. Let's take a look at some Compressor settings:

Threshold

The Threshold is the point at which the Compressor activates. Below that level, it does nothing. Any peaks that exceed the Threshold will go to the next stage for processing, where the peaks will get reduced by an amount you set (so that the overall volume can be increased). That occurs in the next stage and is controlled by the Ratio.

Ratio

The *Ratio* is how much the Compressor is going to reduce the peaks once it's activated.

A 2:1 Ratio setting means the Compressor will lower the level of any peak to one-half of its original value. (But that reduction only applies to that portion of the signal that's above the Threshold.)

The higher the Ratio, the more volume reduction it can do. But at very high Ratio settings, the overall result can sound unnatural. A 2:1, 3:1, or even 4:1 Ratio will work with narration tracks in most cases. But a 10:1 Ratio will make tracks sound weird and "overprocessed." A Ratio of 1:1 reduces the peak to 100% of its original value, which is no reduction at all.

For most voiceovers and talking heads, if you are aiming for an average audio level between -15 dB and -10 dB, and if your absolute Do Not Exceed target is 0 dB, you only have 10 dB of headroom to work with. So you might set the Threshold at -10 and the Ratio at 3:1. In that case, everything below -10 dB would be left alone. Once that line has been crossed, however, everything above it would be reduced to 30% of its level. So a 0 dB peak would get trimmed to -7 dB. That's slightly above the -10 dB target, but it is still less than 0 dB, preserves the dynamic range, and it's not clipping or splattering either.

More Examples

If the Threshold is set to -10 dB, and the Ratio is 2:1, and the peak is a whopping +20 dB, here's the result:

The Compressor will activate because the peak is above the Threshold. The reduction Ratio is 2:1, so the level above the Threshold will be reduced to 50% of its original value. Since the peak is 30 dB above *the Threshold,* the peak will be reduced to 50% of 30 dB (or 15 dB) *above the Threshold,* which is +5 dB in absolute terms. That's a reduction, alright, but the red line still gets crossed because the peak is +5 dB. Changing the Threshold won't help because it's activated at this level already.

What happens if you turn the Ratio up? At 3:1, the result would have the peak hit exactly at 0 dB, which doesn't shoot through the ceiling (technically) but does touch it and allows no room for error. The Ratio could be increased to 4:1 or even beyond, but higher Ratio levels start to add unwanted compression artifacts—particularly on vocal tracks.

A better way would be to use a lower Ratio (2:1 or 3:1), and in addition to the compressor, turn on a *Limiter* so that no matter what, the 0 dB red line will never be exceeded. Using a Limiter in conjunction with a Compressor means you don't have to be as aggressive with the Compressor settings.

Limiter

Why not just use a Limiter and be done with it? A Limiter grabs whatever exceeds the Threshold setting and lowers the level to the red line value or below. It doesn't let anything get in its way. But it's a chain saw when all you needed was a haircut.

It's OK, even desirable, as a fail-safe, because the audible effect of a Limiter occasionally killing a peak is better than ever exceeding it. But banging the Limiter on every syllable is quite noticeable. And in any case, when used with a compressor, there will be a lot fewer peaks the Limiter will have to take care of. So in crafting a clean, clear, powerful sound, using both a Limiter and a Compressor is the way to go. And really, a Limiter is really a Compressor set to a very high ratio. High ratios are audible, so always use a Compressor before the Limiter so the Limiter will have less to do.

Knobs and Dials

Compressors and Limiters have several other settings you need to know about (depending on the plugin).

Input and Output

Input and Output are just volume controls that set the input and output volume. The Input should (obviously) be set high, but not high enough to clip. The output can be cranked up if the Compressor is controlling the peaks.

When setting levels in an audio chain, what you don't want to do is have your controls fighting each other. Don't turn one way up just to have to turn the next one way down.

Attack and Release

It takes a finite about of time for the Compressor to recognize that a peak is going to exceed the Threshold setting. How quickly it reacts is set by Attack.

Once the Compressor has activated (and lowers the volume by the amount set by Ratio), the next question is, "How long should it stay active?" You might think it's a good idea to deactivate the Compressor quickly. That can work. But in some cases, deactivating too quickly can be heard. Typically, you want to do that smoothly—a dissolve rather than a cut, if you will.

Think about a cymbal. If the Compressor was on the crash cymbal, you might want to activate a bit on the slow side to let some of the transients through. Or you might want it to activate very fast in order to tame the transients. Ditto with the release time. In either case, you'd have to listen to it.

All musical instruments have characteristics that make certain Attack and Decay settings desirable. For some, you want a slower Attack and a faster Release. For others, just the opposite. But you have to listen. I know quite a few audio engineers here in Nashville, and none of them works by rote or by the book: all of them adjust by *ear*.

Vocals are a special case. Usually, with vocals. we want a fairly rapid Attack and a somewhat slower Release. Again, you'll have to listen. The best way is to loop a representative section of a clip and adjust these settings until you can (1) achieve your goal of loudness without exceeding the red line of 0 dB and (2) without anyone being able to hear what you've done (no artifacts). You can usually see with your eyes (on the meters) if you have achieved enough loudness without clipping, but you can only tell you have done it with no audible artifacts if you listen carefully.

When the Compressor is set to rather modest levels like the Threshold of -10 dB and the Ratio of 3:1, you might want both the Attack and Release to occur rather fast. It may be better to just get it over with quickly. And the Attack and Release timing probably won't matter very much at low compression settings. Attack and Release times become more important as more aggressive Compressor settings start becoming audible.

Safe Harbor

For standard announcer/presenter/voiceover tracks, it's usually safe to set the compressor's Threshold at -10 dB and the Ratio at 3:1 (assuming you are aiming for a -15 to -10 dB average). Set the Attack to a fairly fast setting and the Release to a fairly slow setting. These settings should allow you to turn the volume up without splatter if the track was recorded properly to begin with.

Limiter Settings

The Limiter is easy to set. The horizontal blue line is the Do Not Exceed threshold. Usually, that's shoved almost all the way to 0 dB (almost). For safety, set the Threshold to something just under 0 dB. Set *Make Up* to something around -5. Then watch the Gain Reduction display while you play the track. You only want to see the blue indicator flash occasionally (meaning limiting is taking place no more than once every few seconds or so). If it's more than that, turn the Threshold up or use more compression.

Red Bars

There are two red bars immediately above the VU displays on the Mixer. If the audio *ever exceeds* 0 dB, one (or both) of these red bars will light up and will stay lit long enough for you to notice. If that happens, back off the faders till it doesn't.

Loudness Meter

Resolve has a Loudness Meter in Fairlight. It's designed to help producers and broadcasters meet the loudness standards of the FCC and other regulatory authorities. While these standards are not applicable to the Internet, you can monitor the Loudness Meter and see the effect of volume controls, as well as compression and limiting. The average loudness of a track involves a time element. The right meter uses a shorter time scale. The left one, a longer time scale.

LoudMax

Fairlight's plugins are very high quality, but the selection is limited. Fortunately, Resolve also allows you to use VST plugins (Effects > VST Effects), many of which are available for free. One you should be aware of is LoudMax. It acts very much like a Compressor/Limiter but with a much simpler control scheme and produces clear, loud audio tracks. Even though it's free, it's better than any I have bought.

You can get it at https://loudmax.blogspot.com (and if you do, leave a little donation on the site for the folks who developed it).

LoudMax greatly simplifies the whole Compressor/Limiter thing. In a nutshell, there are two faders. The top control is the Threshold. Around -9 dB is a good starting point. The bottom control sets output level (basically, "Ratio"). -3 dB is a reasonable place to start. But play the Track and adjust the output level such that you are hitting around -10 dB to -5 dB and never touching 0 dB on your meters with the Track's fader set to -10. In any case, there are only two controls, and it's easy to hear what they are doing to the Track. And like everything else in the universe, there are YouTube videos to help you adjust LoudMax should you need them. David Harry has a very nice video on YouTube to help with this plug-in. (He also has links to download the plug-in.)

Normalizing

Another approach to evening out an audio track is to "normalize" the audio. This is nothing more than boosting or reducing the level of the selected track so that no peak exceeds a specific dB. During playback, if the audio is peaking around -11 dB, you could normalize by setting the level to 6 dB (ending up with peaks at -5 dB). To access this function, select a clip, right-click, then *Normalize Audio Levels*.

However, normalizing a track raises or lowers the volume of the entire track until *the peaks* match the setting desired. Usually, normalizing lowers the volume of the entire track so that no peak exceeds the level specified. The Compressor only works on the peaks themselves.

Another way, of course, to tame a few errant peaks is to use keyframes in the audio track. It's easy to find errant peaks by simply looking at the audio waveform. If you notice a peak you want to tame, add several keyframes (typically 5) around the peak, and drop the level of the center keyframe a few dB (forming a smooth (inverted) bell curve so what you just did won't be audible). Another method is to edit in a snippet from the safety track recorded at a lower volume. Which technique you use depends on how many peaks you need to tame and the nature of the recording itself. Compressor/Limiters (such as FatMax) are commonly used on every vocal Track, while the keyframe/snippet method is used to remove occasional pops and mic bumps.

Finally, note that effects added via Fairlight are not indicated on the clip in the Timeline while Fairlight (and other) effects added via the Edit page are (via the FX icon bottom left).

With Audio Effects, Less is More

Over the years, I've spent a lot of time in control rooms recording audio for video. I've learned the hard way that too much of a good thing is too much.

One of the biggest television events of all time was *Roots*, based on the best-selling book by Alex Haley. ICYDK, it's the story of a Black American tracing his roots back to Africa. It was and still is a helluva story. The book sold 1.5 million copies *in hardback* and won the Pulitzer Prize. In the miniseries, Haley was played by James Earl Jones. It's an iconic performance. The miniseries won nine Emmys and a Peabody. Not only was Alex a cultural icon, he was a national treasure.

Some years later, I was working with Alex on a documentary. I was writing/producing, and he was narrating. When I wrote the script, the voice I had in my head was the voice of James Earl Jones—you know, the guy who voices *Darth Vader*. So I wrote the script with grandiose verbiage (read: pretentious). The kind of words that needed to be read in tremendous stentorian tones. But when it came time to record the tracks, Alex Haley showed up, not James Earl Jones. And Alex was a very mild-mannered, soft-spoken guy. Those words just didn't fit his voice. At all.

When we were done, I took the tape (honest to god reel-to-reel in those days), cut it up into individual words and phrases, and taped all these snippets to the wall (hundreds and hundreds of them). Over the next several weeks, I painstakingly listened to every word and selected the most "James Earl Jones-sounding" version in building the master track. But that wasn't enough. It needed more.

About that time, the Eventide Harmonizer had been released that would let you change the pitch without affecting the timing (with Resolve, you can do this via the Inspector). I ran the track through the Eventide and took the vocal pitch down. Way down. Then I used parametric EQ, flat-plate reverb, automatic double tracking—basically every trick in the book. Weeks later (yes, weeks), I finally finished the audio track. And if I do say so myself, by the time I locked the track, Alex Haley actually sounded quite a bit like James Earl Jones.

The music was by Aaron Copeland, which added even more gravitas. And for the documentary's premier, I had a sound system installed in the theater that was specifically designed to complement his voice. An audio technician tuned the hall to make Alex as James Earl Jones-sounding as the technology would allow.

On the night of the premiere, Alex shows up in his tux. We exchanged pleasantries. Then I realized that he had not heard the track. He had been in LA most of the time and never got to the studio to review and approve it. But I had so distorted his voice that he, Alex Haley, was no longer Alex Haley. I had a written release that covered that sort of thing, but this could get personal and ugly. What would he think? What would he say? What would he do?

The lights dimmed, and Appalachian Spring (fourth movement) thundered from an entire wall of Klipsch La Scala monitors. Then the narration started.

I couldn't look at Alex. My shoes were literally filling with sweat. Finally, thank God, it was over. What should have been a triumph was instead a total, unmitigated disaster. Then I saw Alex walking toward me, smiling broadly. At that moment, I realized that in *his own head*, Alex's internal voice is also that of James Earl Jones. He thought narration sounded *exactly* like him.

The Deliver Page

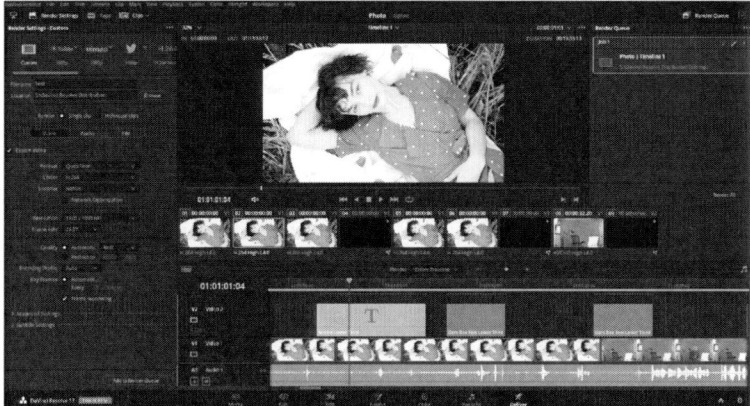

There's an old saying that "There's many a slip 'twixt the cup and the lip." You've taken pains with the lighting, and you've spent days and days shooting, editing, polishing, and grading your video. The audio track is pristine. Now it's time to upload it to YouTube, Vimeo, Twitter, Facebook, Instagram, or some other social media platform or CDN.

When they get your file, they are going to decompress it, fiddle with it, and recompress it again. You don't have much choice in how any of that is done. But in order to get your file to them, you'll first have to compress and encode it. Resolve calls this step a "final render." This is your very last chance to affect the quality of your project. And it can't improve on what you've done—it can only degrade it, so you want it done as carefully as possible.

Video Compression

Compression is a necessary evil because virtually no one has the bandwidth to watch your video in real-time via the Internet unless it has been compressed. Netflix is compressed. So are Hulu and Disney, and everything on Netflix, Dish, Amazon, and cable—*they all do it.*

With compression, you are trading off some loss in quality for the ability to reach a vastly wider audience.

To make that deal work in your favor, you want to sacrifice as little quality as you can and still gain the reach of a large platform like YouTube, Vimeo, other social media, CDNs, and more importantly, their viewers who have varying amounts of bandwidth. Fortunately, video contains a lot of redundant information that can be removed or restated in a smaller form and which can be used to reconstruct it rather well on the other side. These days, compressed video looks remarkably good, considering how much of the original was removed, repackaged, and reconstructed.

The Deliver page is where the final product gets rendered, compressed, and wrapped for delivery. Basically, you select a compression and encoding scheme (codec) via *Render Settings* and provide the file name and folder location. The project is put into the *Render Queue*, where nothing happens until you click *Render All*. Then Resolve does its thing. When it's done, Resolve will have created an output file and stored it where you told it to, or, if you asked, will have uploaded the file to the platform of your choice (if your login credentials have been setup in Preferences > System > Internet Accounts).

This two-part process first performs a very thorough final render and then compresses the resulting file (a copy of which is stored locally even if the final result is also uploaded to the cloud). It's going to take a while.

Quick Export

For most users with straightforward projects headed for YouTube, Vimeo, or Twitter, the Quick Export option is faster than using the Deliver page (File > *Quick Export*). This option is available from the Cut, Edit, Fusion, and Color pages.

Quick Export gives you all of the standard presets—H.264 or H.265–as well as presets for YouTube, Vimeo, Twitter (and ProRes if you are on a Mac).

Quick Export has a fixed set of presets, and you can't change them. If you need to do that, you'll have to work from the Deliver page, where you can create your own presets from scratch or modify existing ones.

One advantage of using the Deliver page is that you can queue up multiple renders and let Resolve crank on all of them while you are doing other things. Another reason to use the Deliver page would be if you had to use the Custom settings because the presets just won't do (because the *quality* of the final product as distributed by the platform (CDN or Social Media) is not to your liking).

Another reason for outputting from the Deliver page is that you can set I&O points in the project's Timeline and export just a small portion for testing and verification. If you created a mobile version, for example, and just want to make sure it works on smartphones, you could output a 10-second clip and test it. That's a lot faster than outputting the entire program over and over while you experiment with various settings.

Deliver Page

The first thing to do when you hit the Deliver page is to click on the Sidebar control at the extreme top left and top right to expand the controls areas of the screen (you probably don't need to see as much of the Timeline at this point).

> **NOTE**: There is a light-gray slider under the Render Settings preset icons which reveals the complete list of presets available to you (not just the ones from Quick Export).

Options

On the Deliver page, you have essentially four options:

(1) Render and upload to directly to YouTube, Vimeo, or Twitter using the preset settings.

(2) Use Custom, H.264, H.265, or ProRes (Mac only) and upload the resulting file to YouTube, Vimeo, or Twitter having changed the factory presets to your liking.

(3) Render via Custom, H.264. H.265 or ProRes (Mac only) and upload the resulting file to your company's platform, a WordPress host, or the server in your basement. You can, of course, use Custom, H.265/265, or ProRes presets and upload the resulting file (wherever you put it) to the YouTube, Vimeo, Twitter or other platforms.

(4) Output just a short piece of a project for testing purposes. This will create a file somewhere.

The first option is, of course, the easiest. One click, and you're done. (Well, two clicks if you include Render All.) However, if that's the case (and you don't need to queue multiple projects to render overnight), just use Quick Export instead. It appears that most YouTube videos uploaded from Resolve use the default Render Settings. At least try them out before having to get your mind warped having to learn how video compression works.

The second option is when you want to have a lot more say in how your video gets handled by YouTube, Vimeo, and other platforms than the presets allow. Unless you have unusual quality or source requirements, using the YouTube, Vimeo, or Twitter standard/default presets in Resolve 18, even with their limited range of options, should be fine in most circumstances. However, like everything in video and audio, the proof is in the viewing/listening. Generally speaking, the purpose of the various settings is to ensure that you are uploading the highest quality version of your final edit.

The third option–rolling your own–is problematic in two ways. First, you'll have to get very involved with how your server/host handles video. To make this work, you'll have to figure out the best values for more than two dozen encoder settings (see below). And unless you really learn about video compression techniques, the result is unlikely to be better than Resolve's standard Render Settings. And YouTube, Vimeo, and Twitter also handle a lot of this for you. For these reasons, with WordPress and similar platforms, it's usually best to upload to YouTube or Vimeo and embed *their player* into your WordPress site rather than letting your WP host serve the video.

The fourth option allows you to set I&O points and just output a snippet for review. This is especially helpful for looking at how a problematic scene is going to look or for checking a long-form project before starting the final render and going home for the night.

Workflow

The Deliver page workflow is simple. Select a Render Settings preset or Custom. Enter a filename and the location where you want your local copy archived. Make any changes to the presets you deem necessary. Click "Add to Render Queue." The project will be queued as a job in Render Queue (right side). Then when you are ready, and the kettle is on the boil, click "Render All." When finished, the resulting file will be stored where you told Resolve to put it, as well as uploaded to YouTube, Vimeo, or Twitter if you checked the box. I strongly advise reviewing the final product on the platform once they've confirmed it's uploaded to make sure nothing got lost in translation.

Settings

It is unlikely your knob twisting will improve on the Deliver presets without a deep understanding of what these controls do and how to use them. I highly recommend Casey Faris' "Best YouTube render settings in Resolve 17" (which still apply to Resolve 18). He actually went to the trouble of testing dozens of parameters to determine, once and for all, what the optimal settings should be. And Resolve 18 doesn't change that advice. But if you want to roll your own, here's what the controls do, along with suggested starting points:

Resolution: With platforms that deliver adaptive bitrate streaming (YouTube and most others these days), you should upload using the highest resolution and bitrate the platform will allow. When they stream your video, the platform will lower the resolution/bitrate to match the bandwidth each viewer has available at that moment. On the other hand, if you have to create a file for your company's server or one that has to fit on a thumb drive, you might need to output at some lower bitrate or resolution to accommodate fixed user bandwidth or file size restrictions.

Frame rate: Typically, this is the frame rate of the Timeline. For most social media platforms, it should be 30 FPS.

> **CAVEAT**: If you set the Resolution on the Deliver page higher than the Timeline's actual resolution, *Resolve will upscale your project* from the lower res Timeline. Normally, that's a very bad idea. With 4K footage and an HD Timeline, for example, if you output at 4K from the *Deliver* page, Resolve will *upscale* your HD Timeline rather than using the 4K source footage. The result is not nearly the same as outputting 4K from a 4K Timeline. If this is your workflow, *change the Timeline resolution to 4K before you hit the Deliver page* and deal with that change first. However, there's been a lot of grumbling amongst the masses about this "feature," so how it works may change in future versions.

Format: Your choices for YouTube and Vimeo are QuickTime or MP4. Resolve defaults to MP4, which is YouTube's preference. On the other hand, YouTube supports .mov, so my choice would be QuickTime for both.

Codec: Here, codec means the *distribution* codec. For Windows, H.264 was long the standard for HD, while H.265 was developed for 4K and up. I now use H.265 for everything. In fact, the consensus seems to be building for H.265 to be the new standard. While H.264/265 are not ideal for acquisition or editing (for reasons already covered), these codecs are fine for distribution. You could use a DNx or ProRes codec at this point, but the problems often encountered when editing with interframe codecs like H.264/265 are not relevant for the distribution file. With interframe codecs, it may take a bit longer for YouTube or Vimeo to unpack your file and recompress it, but the final quality will be similar, if not identical. For this reason, using H.264 or H.265 (preferred) as the upload codec will be fine unless you have special requirements.

For Mac Users

Apple ProRes offers many more choices of what is probably the best codec available (which is why I have sometimes use an older MacBook to encode in ProRes).

422 Proxy: At the bottom of the ProRes hierarchy is 422 Proxy. This codec is not really intended for distribution (it's for offline workflows).

422 LT: The next step up is 422 LT, the lite version of ProRes. Even so, the target (assumed) bit rate is 108 Mbps. Very good results and the smallest file of the distributable ProRes versions.

422: ProRes 422 has been the standard for years. Supports 10-bit 4:2:2 video with a bit rate of 147 Mbps. In fact, there's no point going beyond 422 unless the camera/footage/timeline actually produces images that require higher bit rates or needs alpha channel support. *Probably the best all-around choice for a distribution codec.*

422 HQ: Essentially 422 with a higher bit rate of 220 Mbps. While this is Resolve's default setting, using 422 will reduce the file size by quite a bit without noticeably lowering the quality. While 422 HQ is marginally higher quality, it would take a very discerning eye to see it. For that reason, 422 is probably the better choice.

4444: Near the pinnacle is ProRes 4444. Supports RGB color space with a target data rate of 330 Mbps and 12-bit color depth. Probably overkill unless your project is very high-end and your camera and Timeline support it. Sometimes used for animation. Supports alpha channel. Not normally used for distribution.

4444 XQ: If you care enough to send the very best, 4444 XQ checks in at a whopping 500 Mbps. Unless your camera is an Arri Alexa, this setting is not going to be useful. Not normally used for distribution.

Type: This is the type of encoder Resolve is going to use. For H.264 or H.265, if your graphics card has a *hardware* encoder available (AMD, Nvidia), select it. Hardware encoding makes the final render go much faster. If the platform you are uploading to accepts H.265, use that because it's a faster final render than H.264. You can verify this on Reddit, but I know of no reason not to use H.265 for everything going forward.

Quality: The "Quality" setting on most compression software is measured in by "bit rate," which at one time meant the bandwidth you thought a majority of your viewers had (that was in the olden days). Both YouTube and Vimeo now use *adaptive bit rate*, meaning that when a viewer clicks on a video, the platform determines how much bandwidth that viewer has available to stream at that exact moment, and they adjust the stream's resolution to accommodate the user's bandwidth. You might have uploaded a 4K file, but if a viewer doesn't have the bandwidth to see it in 4K at that exact moment, YouTube and Vimeo will serve them a lower res version.

For historical reasons, bit rate settings for video compression are typically referenced to Kbps (kilobits per second) rather than the Mbps (megabits per second) your broadband provider touts.

With the *Custom* preset, you can control the compressor's settings *for the upload*. Ditto for the H.264 and H.265 *Master* presets. With these, either use *Automatic* set to Best or set *Restrict to* to some incredibly high number. The default is 10,000 Kbps, which is not nothing, but is not really high enough for HD and certainly not enough for 4K and above. I generally use 60,000 Kbps for 1080p HD and up to 80,000 Kbps for 4K.

Casey Faris found (with actual testing) that he got his best results picking a bit rate that was twice the frame rate in Mbps. In other words, for a 30p video, use 60,000 Kbps. And for 24p use 50,000 Kbps. For higher bit rates, Faris didn't see much if any improvement. He did see lots of improvement upscaling the resolution to Ultra HD even if the Timeline is just HD. His theory is that it triggers something at YouTube to get them to handle your file with greater care. IDK...

If you need to reduce the file size (to put on a thumb drive, for example, or to send through an e-mail system), you may need to reduce the bit rate dramatically. Also, the Custom setting does not automatically adjust the bit rate as you change the resolution—you'll have to do that manually. That said, extremely high bit rates really only matter if there's a lot of motion or detail in the video or for higher frame rates.

As mentioned previously, the Vimeo and YouTube presets don't let you change the bit rate from the default value, and the internal value of this setting is not well documented. It could be as low as 10,000 Kbps which is not nearly enough. One option would be to use the H.264 or H.265 Master presets and adjust upward. The file may not upload automatically. In that case, you'll have to do that manually. I use the presets for personal stuff all the time but switch to Master or a Custom preset when dealing with client material. That said, I ran some tests while writing this chapter, and the Vimeo and YouTube presets were actually pretty good for HD.

Encoding Profile: These settings have little effect on quality but can affect file size and how long encoding will take (efficiency). The consensus seems to be that High achieves smaller file sizes but needs the most computing power (and time). Main is the old standard and uses the same encoder as High. Baseline uses a different encoding scheme and seems to be used primarily to compress video for mobile devices. I use High for everything. Another option is to leave it on Auto and let Resolve's AI figure it out based on the project's resolution (it's pretty good at this).

Key Frames: I'd leave this on Auto and let Resolve's AI figure it out.

Audio: This is the source for your audio. Bus 1 (Stereo) is the default. Pick that one.

Audio Codec: The choices are AAC or Linear PCM. Linear PCM typically results in a larger file size, but AAC, despite being lossy, is fine. In any case, YouTube and Vimeo are going to re-encode it to something else anyway.

YouTube publishes its recommended compressor/encoder setting at

SUPPORT.GOOGLE.COM/YOUTUBE/ANSWER/1722171

Vimeo publishes its recommended compressor/encoder setting at

VIMEO.COM/HELP/COMPRESSION

Tim Yemmax has a short video on YouTube with very practical Deliver page settings if you prefer to see these settings in action, and I've mentioned the highly informative Casey Faris' test video already.

Check YouTube and Vimeo's documentation, the Resolve forum, or current YouTube videos occasionally, as preferences and requirements are subject to change. However, note that Resolve does not give you direct access to some of the settings these platforms prefer (not that it matters).

Save Your Work

Once you have figured out your magic formula, under the three dots menu is a "Save as New Preset" option. Save your preset settings so you can apply them to future projects.

Final Step

The final step is to Add to Render Queue and then (right side) click on Render All. This final render cleans up a multitude of sins as well as shortcuts you took during the edit and can take a while.

Optimized Media/Proxy Media/Render Cache

If you used Optimized Media in your project, you could use those files when you render for final output. You can also use the Proxy Media and Render Cache files when processing the final render, too. That makes the final render much faster because Resolve is able to reuse the previously rendered/optimized files instead of starting from scratch.

However, you would have had to have optimized and rendered with high-quality settings (Preferences) all along. Most of the time, Optimized Media, Proxy Media, and Render Cache are used to create lower-quality proxies to improve playback performance during the edit. So the quality of the files in Optimized Media, Proxy Media, and Render Cache is usually low (intentionally). Using Optimized Media, Proxy Media, or Render Cache for output when they were not intended for that purpose will result in a very-low quality output file.

If the time it takes to render a project in final is an issue, you could use Optimized Media, Proxy Media, and Render Cache set to higher quality settings (such as Original or HQ) so you can reuse them at final. However, that means the Optimized Media and Proxy Media files used during the edit will take longer to render in the Timeline—every time. Generally, this workflow is only used for long-form projects where the final render could take hours without re-using the caches.

Resolve's default is not to use them for final output unless you insist.

Halt and Catch Fire

Using Custom settings for final rendering is problematic. If you can use the available presets as-is or by changing some of the default settings, that's a much easier path. The presets only need information you already know or can easily find out, such as resolution and frame rate.

If you have made a valiant effort to make the presets work, and the quality just isn't up to scratch, then you've no choice but to roll up your sleeves and dig into the arcane and mysterious world of video compression. Custom settings require a lot more information, and getting the right answers can be difficult. Obviously, if you need the flexibility offered by the custom settings and you can accept most of the defaults as is with a small number of changes—and you know what those changes do—then, by all means, use Custom settings.

TL;DR (Video Compression)

Compression is the process of reducing the *size* of a video and/or audio file. What does the *physical size* of the stored file have to do with whether or not it will play at a given resolution and frame rate? Doesn't seem connected, does it?

Here's the background. Your video—the one you've watched a hundred times during the edit–is not a video at all. It's a computer data file. Only when you process it with appropriate software on the other end (via a video player) is video and audio created from it. So the issue really boils down to transmitting a very large data file from point A to point B. And the bigger the file, the more time transmitting that file it will take.

Some years back, before (nearly) everybody had high-speed Internet, to get a video over the Web, you'd have to download the file (via FTP), and only when it was finished could you play it on your computer. It might have taken an hour to download a five-minute video with abysmal resolution and a non-standard (slow) frame rate. Even then, the video had to be highly compressed just the get the file down to a size where it could be downloaded in what was then considered to be a reasonable amount of time.

However, if you can compress a video so that a 30-minute video can be downloaded in 30 minutes or less (given the user's bandwidth), then they can play it *while it downloads*—and that's called *streaming*.

Streaming Example

Let's say your original file, uncompressed, weighs in at 20 gigabytes. That's bytes, so in bits, that works out to 160 gigabits. At an assumed user download speed of 25 megabits per second, that file would take 6,400 seconds or 106 minutes to transmit. If it's a 30-minute video, it would not stream under these conditions (at that download speed). But if you compressed the file further, reducing its size to around 5 gigabytes (~40 gigabits), that file could be transferred in 1,666 seconds or about 27 minutes. So that 30 minute video could be streamed with a little bit of headroom to spare.

The problem is, viewers with more bandwidth would also get this heavily compressed version, while viewers with even less bandwidth would still experience buffering. But heavily compressing a file was the only way to ensure that average viewers would be able to stream it. Quality, resolution, and frame rate were often sacrificed for everyone in order to accommodate the folks with the slowest Internet connections. It was a race to the bottom. The way we finally got around this problem is *adaptive bit rate*.

When a file is transmitted, it's not sent out all at once but in small "packets" of perhaps 1,000 bits. When the packet is successfully received on the other end, the receiving computer sends an acknowledgment, and the next packet is transmitted. By timing the requests for the "next packet," a server can determine the receiving computer's actual connection speed (which can vary with competing traffic).

The adaptive bit rate model looks at how fast each viewer's computer is asking for the next packet of data. If it's very fast, a less compressed version of the file will be sent. However, if a viewer's Internet connection is not fast enough to stream the highest quality video file, a smaller, more-compressed version will be sent instead. And if that one is still too big, an even more-compressed version will be transmitted.

YouTube, Vimeo, and most other CDNs now use adaptive bit rate technology and can adjust the streaming rate to fit a user's bandwidth. A 4K video will stream at 4K if a viewer has the bandwidth for it. If they don't, YouTube will try to send the 1080 version. And if that's no good, the 720—all the way down to 144!

IMPORTANT
This means that if you are uploading to a CDN that uses adaptive bit rate, you only care about *the quality of the uploaded version*. File size be damned. The CDN will take care of user bandwidth issues for you–dynamically. This situation has resulted in a paradigm shift in setting video compression parameters. It used to be you had to pick a sweet spot and compress to the "meaty part" of the bell curve. And there is lots advice online about how to do that. But those recommendations are out of date if your CDN is using adaptive bit rate.

Because YouTube, Vimeo and other social media platforms now use adaptive bit rate, you'll want to have the very best/least compressed version of your video on their server. Unfortunately, Resolve's encoder does not know about adaptive bit rate. You can only give it a single bit rate setting, and the default is a rather low number.

To fix that, you basically lie to the encoder and tell it your viewers have tons of bandwidth. That causes the encoder to create the least-compressed/highest quality file it can for uploading. The CDN will recompress your high-quality video in a variety of lesser resolutions. Resolve defaults to 10,000 Kbps, while most professionals suggest 60,000 Kbps for 1080 HD and a whopping 80,000 Kbps for 4K or higher.

If you are uploading to a company platform, a WordPress host, or to the server in your basement, and your viewers will have only one file to choose from, you may have to dial down the bit rate (or accept the default) to accommodate viewers with limited bandwidth (assuming you care).

You do that by lowering the bit rate setting, which has the effect of making the resulting file smaller. That also reduces quality and adds more artifacts. Or you can upload the file in very high quality with a large file size, in which case viewers with lots of bandwidth will not have a problem streaming while viewers with limited bandwidth will. This is another reason for hosting your videos on YouTube, Vimeo, or some other CDN with adaptive bit rate technology rather than on your own WP server without it.

> **CAVEAT**: If you are reviewing an uploaded video on a CDN that uses adaptive bit rate, your download bandwidth will determine which version you are seeing. If you have limited bandwidth, the CDN will send you a dumbed-down version. If you want to see what these various versions look like, in the case of YouTube and Vimeo, when reviewing a video, open the player's *Settings* cog wheel, and set it one of the resolutions listed instead of Auto.

Stray Cache

When you finish a project, delete the project's caches (Playback > Delete Render Cache > All). If you used Optimized Media or Proxy Media, delete those files as well. These files are huge, and if you edit a lot of projects, your cache drive will fill up fast (and get very sluggish). From time to time, open your cache folder from your OS and look for stray caches and delete them.

Backups and Archiving

Backups and Archiving

If this is the first time you've used DaVinci Resolve 18, you won't have any problem locating your project files, and your computer won't be clogged with bits and pieces of old projects. However, it won't take long for that to change. Thankfully, DaVinci Resolve has several systems that will help you keep the clutter to a minimum and help you organize the amazing number of files that get created even for a small project.

Project Management

When you start a new project, DaVinci Resolve creates a Project file. A *Project* consists of the Timeline plus all of the work you will do to modify the media (video, sound effects, music, B-roll, stills, graphics, etc.). Because of the large size of the files, media is not included in a Project file—only the instructions needed to convert the media into the finished video. Every dissolve, every grade, every Node—every change you made to the media—is included in the Project. But not the media itself. This means that a Project contains the Timeline and everything else that's needed to construct the project *except for the actual media needed to do it*. So when backing up a project, you need to be aware of what you are backing up and what you are not...

Project Libraries

With previous versions, Projects were stored in a "database." With Resolve 18, BMD now calls the databases *Project Libraries*. However, if you have ever installed previous versions of Resolve, your UI may still call them databases. Or Resolve may use both terms indiscriminately. And BMD fiddles with these functions with every release, so by the time you read this, the version of Resolve you have may work somewhat differently (or even have different names for things).

The Timeline and associated files will have been saved many times during the edit as a Project (.drp). After rendering in Deliver, a final output file will also have been created. You'll still have various media—video files, music, stills, and VO tracks (typically) on a drive somewhere. Resolve keeps track of what has happened to all the media in a *Project*, and it keeps track of multiple Projects through *Project Libraries*. Project Libraries is a collection of Projects the same way a library is a collection of books.

To access a Project , go to File > Project Manager. You are given a choice of where Resolve should look for a project: Local (your computer), Network (other computers attached to yours via a computer network), or Cloud (a storage device somewhere in cyberspace). If you are a beginning user, my guess is you'll only have Local libraries. Cloud is new with Resolve 18. Basically, you have to have a Blackmagic Designs (BMD) account (free) and sign up for the Cloud and pay a small monthly fee. In fact, BMD now has a slew of various Cloud options and more on the way as the industry (all industries, really) move more and more stuff to "the cloud."

Local Library/Database

When you ran Resolve 18 for the first time, a *Local Project Library/Database* was created. BMD recommends that you create additional databases as you need them—a new one each year perhaps, or maybe individual databases for large clients plus one for the unwashed masses. The Local Project Libraries are actually stored in the Library on Macs and in the Program Data folder on Windows machines (sometimes). But whatever you do, do not fiddle with any Resolve libraries/databases using your OS. Only use Resolve's library/database management tools to create, backup, archive, or delete them. Although you can store a library/database on any (non-removable) drive connected to your machine, it's a good idea to keep them on the same drive as the Resolve app.

To find out where a library/database is stored, go to File > Project Manager, click on the sidebar icon (next to Projects). Select a library/database and right-click on it. Then select "Open File Location." This folder will be buried somewhere deep on a local drive (as in

Resolve Disk Database > Resolve Projects > Users > guest > Projects). The actual storage locations varies depending on which version of Resolve you are running and whether or not you ever ran a previous version on the same computer. However, as mentioned, don't mess with a library/database file using your OS.

Creating a New Database (Windows and Mac)
This is another one of those annoying quirks where you have to create a folder before Resolve will let you access it. To create a new Project Library/database, use your OS to create a folder on the same drive as the app and give the folder a name like "Resolve Project Libraries."

Under File > Project Manager, click on Add Project Library (bottom button). Give the library a name such as "Sterling Cooper 2022." Then click Location *Browse* and navigate to the folder you created. The last step is *Create* the Project Library. (Not the most elegant nav tool on the planet, but the Resolve 18 version is much better than previous.) Now, when you create a New Project, first identify which Project Library it will reside in. Resolve defaults to last used if you don't pick. BTW, the "house" icon at the lower right of any screen is a shortcut to Project Manager.

Backing up a Project Library
Because a Project (in a Project Library) represents *the entire project* (all the edit/effects work you did) *except for the media itself*, it's important to back up Libraries from time to time, especially if the work is critical. If a Library crashes without a backup, you're screwed. Totally. (Well, you could re-edit the entire project.)

To backup a Library, Export it to some safe place, and, if your working copy is damaged, you can Import the good copy.

Backing up Media
Open the project. The go to File > Media Management (*not Project Manager*). Media Management will let you make a complete copy of *the media* in an Entire Project or just the media used in the Timeline or just specific Clips. But copying an Entire Project this way still does

not capture everything. It only copies the media used in a project and not the Project itself. You can also backup just the media you actually used in the project (Used Media), which saves space. Or you can click on individual clips and then go to Media Management and backup just those clips.

Backing up the Entire Project (Archiving)

Archiving is the only guaranteed way of making sure you've got every little thing needed to reconstruct the project on your computer (or a different one). Before attempting to archive a project, determine where the project's archive folder will be stored. The size of the archive depends on the size of the media in project (as well as caches and Optimized Media/Proxy Media files if used). This folder, since it will contain *everything* needed to reconstruct the project, can be quite large. An external drive or other removable media (or the cloud) is typically used for this purpose.

NOTE that Exporting Project and Exporting Project Archive are not the same thing. Exporting a Project from the File menu just exports the Project file (library/database). Exporting a Project Archive, on the other hand, exports everything, including media.

Go to File > Project Manager. Right-click on the project you want to archive and select *Export Project Archive*. Resolve creates a file with a .dra extension and archives it to an external drive (or whatever storage device you selected).

Then, move across state lines, change computers, plug in the external drive, go to the Media page and File > Project Manager. Right-click on any *empty space* in the Projects panel and select *Restore Project Archive*. Find the project's .dra file and restore. It's a good idea to practice this process a few times before your life (or job) depends on it.

It's very important that you practice archiving and restoring a project to ensure that it actually works. A hardware engineer who worked for me was meticulous about backing up our hard drives, even to the point of keeping them in fire-proof containers and storing them off-site. Then, one day when everything crashed, he was unable to restore the files because some internal software glitch that had occurred *several months earlier,* and the backups were garbage. Turned out he had never actually tried to restore the files from the backups. Oops.

Deleting a Project

To delete a project, go to File > Project Manager. Right-click on the project and select *Delete*. AFAIK, you have to have at least one project in the Project Libraries. Of course, the project's files (or at least remnants) will remain on your storage devices until overwritten.

Last Word

When you finish a project, delete the caches (Playback > Delete Render Cache > All). If you used Optimized Media or Proxy Media, delete those files as well. It used to be you could do that with one click (Playback > Delete Optimized Media), but for some reason, that button is missing from Resolve 18. For the time being, you'll have to manually delete Optimized Media/Proxy Media files using the standard file deletion tools in your OS. Navigate to your cache drive and from there to Davinci Resolve Cache > CacheClip > OptimizedMedia. Ditto for Proxy Media files.

From time to time, navigate to your cache drive and clean out any errant Optimized Media, Proxy Media, and Render Cache files you no longer need. These files are huge, and if you don't delete these files regularly, your cache drive will fill up fast.

Resources

Resources

lackmagic Design has several videos on YouTube, including a 16-minute tutorial for beginners and one where you can learn color grading in only five minutes. Ahem.

Much better is *Blackmagic Design Training* (on their website) with much more comprehensive tutorials along with sample videos to practice with. You can access it from the Resolve 18 Help menu.

The DaVinci Resolve Introduction to Editing course is a great place to start. Not only will it help you use Resolve, it demonstrates proper editing workflows. BMD also has several full-length books online for free. They are rather large pdfs, but they are useful references. Make sure and get a copy of the official *DaVinci Resolve 18 Reference Manual* (3,981 pages) available only as a pdf–*if you have room for it on your hard drive.*

There's also a "Beginners Guide" that's nicely done (but focuses too much on the Cut page, IMO). But it's also a free pdf, and it's much shorter (only 444 pages). You should get both of them. Note that, AFAIK, all BMD books, guides, and manuals are available as free PDF downloads.

BMD hosts a DaVinci Resolve Forum (on their website). It mostly covers bugs and problems. There is also a DaVinci Resolve Reddit as well as other forums too. Good info if you get stuck. However, keep in mind that many of the problems and bugs addressed in the forums and on Reddit were disposed of long ago. Look for the most up-to-date information you can find, or you'll be in the Matrix.

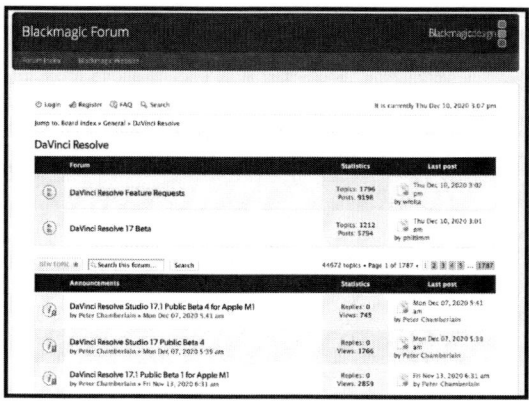

YouTube has hundreds of videos about using various Resolve effects and techniques. In some cases, you may need to watch several videos to figure out how to make them work. However, few of these videos give step-by-step or click-by-click instructions and virtually all expect you to know your way around Resolve to some degree. It's my belief that once you can import video, edit it, apply a few simple effects from the Edit, Color, and Fusion pages, and output a final product that will play on YouTube or some other CDN, you should be able to follow just about any YouTube explainer.

> **CAVEAT**: There are hundreds and hundreds of Tutorials on YouTube. When searching, turn the YouTube filter on and only look for videos made in the past year. Many of the older videos are for earlier versions of Resolve and may not apply to or even work with Resolve 17/18. Also, Filmora is a competing consumer NLE that markets its wares by purporting to offer how-to advice about using Resolve, then segueing into a pitch for Filmora. (I know, right?) Avoid getting Resolve advice from Filmora.

Hardware

Davinci Resolve 17 Benchmarks! Did I get what ...

John's Films is probably the leading YouTube channel for all things related to the hardware needed for Resolve. John is an IT guy who took up filmmaking, so he knows what needs to be in the box. If you want to upgrade your PC or are in the market for a new one, you should definitely visit this channel. I used his videos to tune my new Windows machine when it didn't work as expected, and the performance increase was amazing (and free). (But do buy him a cup of coffee.)

Gerald Undone also has outstandingly thorough video production hardware videos. I always watch his (and John's Films) reviews before buying (and their setup guides after I can't get it to work–beats RTFM).

Video Production

Resolve 17 Fusion - Mac Mini M1 8gb

Filmmaker Central covers audio and video production, lighting, sound, and post. Very grounded, absolutely sound information.

Overview Demo of New Features in DaVinci Resolve...

Nofilmschool doesn't focus exclusively on Resolve, but it does have an extensive collection on anything and everything related to making film and video (focusing more on production than post). Useful even if you didn't go to USC with George Lucas—hence the name. Very informative if you are in the market for new gear.

Switching to Davinci Resolve for 30 DAYS

Moviola.com I mentioned earlier. They have a video tutorial that's not quite as good as attending USC, but it's pretty damned comprehensive. If you are serious about long-form video, moviemaking, documentaries, and the like, this is the place.

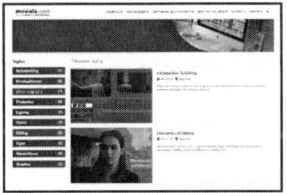

Film Alliance has lots of videos with really clever techniques that will let your gear punch well above its weight. If you want to achieve a really great cinema look, look no further. It's like MacGyver and George Lucas had a baby.

RESOURCES 261

Tutorials

FREE YOUTUBE EDITING MINI-COURSE - Download...

Casey Faris probably has the most viewers of any Resolve channel on YouTube. There's a reason: great content. Faris has long been one of my favorites too. While watching the tutorials, pay close attention to Faris' workflow and operating practices. Emulate.

How to Get Started with Davinci Resolve: The FIRST...

Jay Lippman has scads of Resolve editing and grading tutorials, as well as videos on post-production in general. Very knowledgeable guy, with practical advice even beginners can follow.

DaVinci Resolve - COMPLETE Tutorial for Beginners!

Justin Brown (Primal Video) has a "complete" 30 minute Resolve tutorial that covers a lot of ground in just a half hour. His channel also has lots of advice, techniques, and tricks specifically for making YouTube videos.

Quick Cinematic Look Tutorial : Davinci Resolve 16

Video Editing in a Minute has hundreds and hundreds of short, to-the-point tutorial videos that are one to two minutes max. Great if you don't want a lecture about Sergei Eisenstein's Odessa steps sequence and just want to know how to do something in Resolve right now.

LEARN DAVINCI RESOLVE 17 IN 30 MINUTES - Editing...

Chris' Tutorials has more than 100 short videos on a variety Resolve topics. Also lots of videos on production techniques and YouTube info. Unlike some videos on YT, Chris's videos don't just demo stuff—it's real training. Worth your time.

You're Adding Contrast Wrong - DaVinci Resolve 16...

Nathan Carter's videos are aimed at new users. I always find him posting something useful.

Darren Mostyn likewise is building quite a library of Resolve 17 vids. **Think Media** also has helpful tutorials aimed at "solo shooters."

Grading and Fusion

DaVinci Resolve 16 Cinematic Color Grading...

Sidney Baker-Green has a low-key approach and easy to follow steps that achieve excellent results.

Davinci Resolve 16 - Beginner to Hero Tutorial

Color Grading Central—the name says it all. Topics range from easy to wow. Videos tend to be short form *cut to the chase* type.

Render In Place DaVinci Resolve 17

JayAreTV (Justin Robinson) is an extremely popular YouTube channel for advanced color grading and Fusion. He also makes and sells templates that animate titles and add stunning production values without you having to do any work. Check them out at jayaretv.com.

DaVinci Resolve 17 New Features Overview - Tutorial...

Waqas Qazi is a professional colorist who can show you some unbelievable tricks with Resolve. If you really want to get good at grading, this is the place to go. I particularly recommend his video on "The Most Common Mistakes Every Beginner Colorist Makes." It would take quite a bit of practice to learn his techniques, but the results are stunning.

Edit Page Fusion Effects: Resolve 17

CB Super has lots of easy-to-follow videos about creating effects in Resolve (via the Edit page as well as Fusion.)

Note that YouTube searches for Davinci Resolve "Fusion" will often turn up "Cache Fusion" videos which are about an unrelated Oracle product.

Advanced Davinci Resolve Tutorials

The 7 Essential Color Grading Styles Every Filmmaker Mu...

Learn Color Grading by Alex Jordan is one of the best resources for advanced DaVinci Resolve color grading techniques as well as basic tutorials. Jordan's pleasant demeanor and smooth style make his training videos a pleasure to watch. He is patient and extremely knowledgeable, yet his tutorials are to the point and designed for the casual user who wants to take Resolve to the next level as well as experts. He also runs **film-simplified.com** which has modestly priced online courses. Some of the courses are free, so you should definitely take advantage of those because they are much more comprehensive than the average YouTube videos (which often skip key points).

How To Save & Reuse YOUR Effects In Davinci Resolve...

Billy Rybka has lots of videos about creating stunning special effects in Resolve. But he also has several Resolve 17 update videos and a 17 part series on Resolve for beginners. (Coincidence? I think not.) Rybka's videos are short, to the point, and cover all necessary details. He knows his tricks.

The Modern Filmmaker videos will not only help you better understand Resolve, they will inspire you. Marcel knows his way around Resolve, and the videos cover just about any production topic you can think of. Marcel's videos are also excellent examples of how video and audio should be done. Check out "Upscale Soft 1080P to CRISPY 4K" which focuses on the noise reduction plug-in. Great info if you only have HD and want to make it look like 4K.

How To Edit In Davinci Resolve 16 For Beginners

Hardware Requirements & Configuration

Hardware Requirements and Configuration

The normal hardware recommendations for DaVinci Resolve 18 seem demanding compared to Adobe Premier Pro or Apple's Final Cut Pro. But these recommendations assume you will be working with very high-end media formats and performing lots of the post-production magic Resolve is capable of. If you are not, your hardware requirements will be less—a lot less. And Resolve 18 runs a lot better on standard PCs and Macs (and better still on the new Apple M-series chips) than previous versions. But you will need to run the Studio version to get the most from your hardware's performance–some of the key settings mentioned here are simply not available in the free version (but many are, so do what you can with what you've got).

Virtually all cameras compress video using proprietary encoding/compression schemes (codecs). Some of those can be difficult (computationally) to play back smoothly in real-time on less powerful PCs and Macs. (There are workarounds, however.) Casual users with inexpensive camcorders could, in theory, need the same (expensive) computer hardware as professionals with high-resolution, high-frame-rate cameras. But that's *in theory.*

The reason anyone needs a super-powerful CPU, a GPU with blazing speed, and lightning-fast drives is *time*. The time it takes to transcode media; the time it takes waiting on a render before the clip will play smoothly. If you can stand the delays for transcoding, and the frequent halts to render (or can work them into your workflow), these breaks are less of a problem. If they inconvenience you, better hardware can, of course, help.

Smooth Playback

When you perform an edit or add an effect and try to play it back, Resolve (and every other NLE) needs to "render" the video files, make thumbnails, process audio tracks, update the UI, and a host of other things, all while trying to play back the Timeline at the actual FPS setting. If your computer is fast enough, Resolve will be able to perform all these background tasks as it plays.

The FPS indicator at the viewer's top left is green if the Timeline is playing back at the correct FPS (as determined by the Timeline's frame rate) and red if it's not. There's also a numerical display that shows the actual playback FPS. Most of the time, it's obvious when playback is not going smoothly.

In some cases, particularly with complex effects, your computer simply won't be able to keep up in real-time. In that case, playback may be in fits and starts as Resolve prioritizes processing files rather than playing them back.

At other times your computer won't be able to play the Timeline at all unless Resolve is allowed to halt and render the Timeline changes completely. When it's finished, the results of the render are stored in a temporary "cache," which is why Blackmagic refers to renders as "caches." A cache is just a stored render.

What most people think of as "rendering" is when Resolve has to halt and render, but really Resolve is rendering all the time. When Resolve renders "on the fly," nobody notices. The key question is, "Can your computer render your Timeline at the same time it's playing?" Hardware requirements here and elsewhere are generally based on building a machine that can render most changes and play them back in real-time and do so more or less smoothly without having to halt and render very often.

Hardware Requirements

Editing digital video means moving very large files around while software performs a lot of mathematical computations using your computer's available hardware. All of that moving around and some of the computations are done by the CPU, where raw speed is important and where more cores are desirable (up to a point).

However, in Resolve, most video processing computations are done on the graphics card, where a high-powered GPU with lots of fast VRAM is desirable. Resolve makes greater use of the computational power of the graphics card's GPU than other NLEs. So if you feel the need to upgrade, the first step would be to get a much faster graphics card or a second graphics card of the same brand and model (Studio version only). That said, with the rewrite of the render engine for Resolve 18, some of the processing burden was moved back onto the CPU. The result is a more balanced approach to managing computer resources.

Most modestly-endowed PCs (especially those suitable for gaming) and most MacBooks of recent vintage can run Resolve (depending on the exact build, the Resolve version, and what you plan to do with it). A supercomputer, certainly a super expensive one, isn't necessary unless you simply can't stand the render breaks or are dealing with high-res/high-frame-rate footage, or are stuck shooting with an interframe codec like H.264/265.

I've used a five-year-old MacBook Pro with Resolve and didn't find its performance all that bad (for a laptop). OTOH, until recently, I wouldn't have bought a new MacBook specifically for editing—for that kind of money, people can build an ass-kicking Windows 10 machine. Battery life, heat, and small size have long been the priorities for Apple, and that works against applications like Resolve that purely needed speed—until now.

These days, Apple is designing and manufacturing its own chips, and the new M1 and M2 series processors are getting great reviews. Beginning with Resolve 17.4, BMD added native support for the M1 chips, and it delivers astounding performance. As of this writing, it's unclear if BMD will need to make similar changes to get top performance from the newer M2 chip, but if so, you can expect that sometime in Resolve 18's lifecycle. You can also expect to pay more for a MacBook Pro with the power to edit 4K footage without proxies and even 8K RED RAW, but, hey, it's a Mac.

While DaVinci Resolve 15 really taxed a modestly endowed PC, versions 16 and 17 added features that were even more demanding. Many of the YouTube videos of that era about using Resolve (and a huge chunk of previous versions of this book) were aimed at getting Resolve to run without stuttering and glitching on "average" computers. With Resolve 18, the situation has improved to the point that most of you should have few problems getting Resolve to play back smoothly on your PC or Mac. If your computer balks when playing clips with lots of transitions and effects added, there's a lot of information in this book, in various forums, and on YouTube about setup and operating practices that can help. But if your PC or Mac can run the new Resolve 18 without stuttering, glitching, or taking forever to render, *you can skip this section* (until you need them).

End Game

At the end of the day, your computer should be capable of playing original un-edited footage from your camera without stuttering or glitching—playing smoothly at the Timeline's actual FPS. That's not really a Resolve issue. Whether or not your camera's original footage plays smoothly is really determined by the resolution, frame rate, and codec your camera uses, as well as your CPU's and GPU's speed and memory.

A more powerful computer can obviously play back the Timeline at the proper FPS more often than a less powerful one, would need to halt and render less often, and can render faster when that's necessary. But Resolve's *settings* are just as important as your camera's codec and the raw power of the machine your footage is being played on.

Hardware Obsession

While there are minimum hardware requirements, and a more powerful graphics card (or cards) will definitely improve playback performance, most any gaming PC will run Resolve 18 adequately for most casual users, as will most any recent MacBook or iMac *if set up properly*. You may need to upgrade your hardware at some point, but there are Resolve settings and workflow practices that give much better performance results than you typically get from incremental hardware upgrades (and they are free).

Whether or not Resolve can play the Timeline with edits and effects smoothly without rendering too often or too long depends more on the type of codec your camera uses to record video, whether or not you are using the free or paid (Studio) version of Resolve 18, and how you have set up Resolve's various preferences than the hardware it's running on. This means that upgrading your hardware should not be the first thing on your list when looking to enhance Resolve's performance.

Friends who do this for a living edit on very powerful custom-built workstations. Once they've added a very complex effect, rendering takes place not on their machine but in a separate room called a "render farm." Basically, that's a room with racks and racks of PCs with powerful CPUs and multiple graphics cards with equally powerful GPUs (and lots of cooling). The workload is spread out over dozens of machines, and everything happens fast because, for them, time is money. That kind of setup is necessary if you are getting paid by HBO or Netflix. But most people working with Resolve do not need computer hardware capable of playing back anything they can think of doing—no matter how complex—in real-time with virtually no rendering.

When I started editing film, we made our edits with scotch tape and a grease pencil and sent the edited workprint off to a lab. It took at least two weeks to get a conformed print back. *Two weeks to "render out" a 20-minute video.* Later, we moved to videotape, and when we finished editing, we ended up with an Edit Decision List or EDL. That list controlled a bank of videotape machines that we set running right before we went home for the day. All night long, the videotape machines clicked and clacked, and by the next morning, we had an edited master to look at (most of the time).

When non-linear desktop editing took off, PCs running Final Cut or Premier Pro were not very powerful, but neither was the software. Still, you had to be judicious about what and when to render. I sometimes had projects on separate iMacs and alternated working on one while the other rendered (two different projects, of course). It wasn't unusual to have the final render take several hours (as in the days of tape, those were often queued to run overnight).

Making similar adjustments to your workflow can speed things up quite a bit. For example, I had a stand-up where, undesirably, a car enters and leaves frame. I wanted to remove the car using Resolve's motion tracking/Object Removal tools. The entire clip was 30 seconds, but the part with the errant car was only five seconds long. So I used the razor blade tool and cut the clip one frame before the car enters and one frame after it leaves—essentially creating three clips. I applied tracking/Object Removal only to the middle five-second snippet instead of the entire 30-second clip. The render time was cut to only a fraction of what it would have taken to process those effects over the entire clip (which would have been totally unnecessary in any case).

How you actually perform editing tasks is often just as important as the hardware you are performing them on. We make these rational tradeoffs all the time. If you are late for work fairly often, you could spend a lot of money on a really fast car (and the speeding tickets that usually accompany fast cars), or you could set your alarm clock a bit earlier. Either method can get you to work on time. Simply changing how you work is often a more practical alternative to buying better hardware and is certainly cheaper.

If you have performance issues and so-so hardware, look for ways you can tweak your workflow and editing process so that you are not waiting for renders all the time and not bothered by them when they occur. By choosing when you render projects and which clips you render, and even which sections of clips you render, you can probably live with less than state-of-the-art hardware. I started editing with Resolve on a MacBook Pro and thought it was fine. I built a Windows 10 machine that has lots more power, but frankly, because of the way I use it, it didn't seem that much of a jump in performance. The cost of incremental performance increases can be high and, in my view, often unnecessary except for digital media professionals.

Hardware Recommendations

It is difficult to give blanket hardware recommendations for Resolve—yet everybody seems to do it. The hardware you need really depends on your camera's resolution, codec, frame rate, and whether or not you are just using Resolve's basic editing features or you're a heavy user of OpenFX effects and Fusion. How long you are willing to wait for an effect to be rendered is also a factor. And while Resolve 18's hardware requirements are less than for previous versions, it's still a beast that does benefit from high-performance hardware.

Your workflow and your camera largely determine how much power your computer hardware needs to have. Some types of work places great demands on the CPU and system memory. Other types of work may need a high-performance GPU (or more than one) or high-capacity drives with very high data rates. Obviously, if you are a professional colorist or edit network TV shows or music videos for a label, you have different hardware requirements than if you are just editing straightforward talking head videos for social media.

Benchmarking

When you are shopping for new hardware, you can find a lot of data on how specific CPUs, GPUs, and other hardware perform on benchmark tests. Most of those tests give results for benchmarks that mean something to gamers. While these scores can be useful, many of the numbers don't apply directly to the demands of video editing. I liken the gamer benchmark tests to engine comparisons in car magazines. Zero to 60 isn't a spec that means much to me driving on public streets with cops around—likewise, with most of the spec comparisons on YouTube. I'm not convinced that I should spend an extra couple of hundred dollars on a card that offers a performance increase based on running a *gaming* benchmark.

Video editing benchmarks are more useful. One popular test suite, Cinebench, is designed to throw off numbers that relate to typical video production tasks rather than games, and it's good for apples-to-apples comparisons between different hardware builds, but the test results may not accurately predict the performance of your hardware when editing *your camera's footage* in your workflow. Puget Systems also has a very good benchmarking suite they also let you download for free. This one runs the wheels off your system and compares it to their top-of-the-line workstation using a variety of consumer, prosumer, and high-end codecs—so don't expect your gear to ace these tests.

```
-----System Specs-----
Computer Name: DESKTOP-4844PQO
CPU: AMD Ryzen 5 3600 6-Core Processor
GPU: Radeon RX 5500 XT (26.20.15029.27017)
RAM: 32GB
Storage: WDC WD2003FZEX-00SRLA0 (S:)-Samsung SSD 970 EVO 500GB (C:)-Samsung SSD 970 EVO Plus 1TB (D:)
OS Version: Windows 10 Pro (Ver. 1909)
---------

Puget Systems DaVinci Resolve Studio Benchmark V0.6 BETA (DR 16.2.5.15),4K,Overall Score,432,Score
Puget Systems DaVinci Resolve Studio Benchmark V0.6 BETA (DR 16.2.5.15),4K Test Average,Basic Grade,53.7,Score
Puget Systems DaVinci Resolve Studio Benchmark V0.6 BETA (DR 16.2.5.15),4K Test Average,OpenFX - Lens Flare + Tilt-Shift Blur + Sharpen,33.5,Score
Puget Systems DaVinci Resolve Studio Benchmark V0.6 BETA (DR 16.2.5.15),4K Test Average,Temporal NR - Better 2 Frames,40.2,Score
Puget Systems DaVinci Resolve Studio Benchmark V0.6 BETA (DR 16.2.5.15),4K Test Average,3x Temporal NR - Better 2 Frames,31.3,Score
Puget Systems DaVinci Resolve Studio Benchmark V0.6 BETA (DR 16.2.5.15),4K Test Average,Optimized Media,57.2,Score
Puget Systems DaVinci Resolve Studio Benchmark V0.6 BETA (DR 16.2.5.15),4K CinemaRAW Light,Codec Average,34.5,Score
Puget Systems DaVinci Resolve Studio Benchmark V0.6 BETA (DR 16.2.5.15),4K H264 150Mbps 8bit,Codec Average,45.8,Score
Puget Systems DaVinci Resolve Studio Benchmark V0.6 BETA (DR 16.2.5.15),4K ProRes 422,Codec Average,42.8,Score
Puget Systems DaVinci Resolve Studio Benchmark V0.6 BETA (DR 16.2.5.15),4K ProRes 4444,Codec Average,49,Score
Puget Systems DaVinci Resolve Studio Benchmark V0.6 BETA (DR 16.2.5.15),4K RED,Codec Average,43.8,Score
Puget Systems DaVinci Resolve Studio Benchmark V0.6 BETA (DR 16.2.5.15),4K CinemaRAW Light,Basic Grade,19.16,FPS
Puget Systems DaVinci Resolve Studio Benchmark V0.6 BETA (DR 16.2.5.15),4K CinemaRAW Light,OpenFX - Lens Flare + Tilt-Shift Blur + Sharpen,5.65,FPS
Puget Systems DaVinci Resolve Studio Benchmark V0.6 BETA (DR 16.2.5.15),4K CinemaRAW Light,Temporal NR - Better 2 Frames,5.98,FPS
Puget Systems DaVinci Resolve Studio Benchmark V0.6 BETA (DR 16.2.5.15),4K CinemaRAW Light,3x Temporal NR - Better 2 Frames,2.06,FPS
Puget Systems DaVinci Resolve Studio Benchmark V0.6 BETA (DR 16.2.5.15),4K CinemaRAW Light,Optimized Media,31.74,FPS
Puget Systems DaVinci Resolve Studio Benchmark V0.6 BETA (DR 16.2.5.15),4K H264 150Mbps 8bit,Basic Grade,25.87,FPS
Puget Systems DaVinci Resolve Studio Benchmark V0.6 BETA (DR 16.2.5.15),4K H264 150Mbps 8bit,OpenFX - Lens Flare + Tilt-Shift Blur + Sharpen,6.34,FPS
Puget Systems DaVinci Resolve Studio Benchmark V0.6 BETA (DR 16.2.5.15),4K H264 150Mbps 8bit,Temporal NR - Better 2 Frames,7.08,FPS
Puget Systems DaVinci Resolve Studio Benchmark V0.6 BETA (DR 16.2.5.15),4K H264 150Mbps 8bit,3x Temporal NR - Better 2 Frames,2.49,FPS
Puget Systems DaVinci Resolve Studio Benchmark V0.6 BETA (DR 16.2.5.15),4K H264 150Mbps 8bit,Optimized Media,58.29,FPS
Puget Systems DaVinci Resolve Studio Benchmark V0.6 BETA (DR 16.2.5.15),4K ProRes 422,Basic Grade,29.48,FPS
Puget Systems DaVinci Resolve Studio Benchmark V0.6 BETA (DR 16.2.5.15),4K ProRes 422,OpenFX - Lens Flare + Tilt-Shift Blur + Sharpen,6.81,FPS
Puget Systems DaVinci Resolve Studio Benchmark V0.6 BETA (DR 16.2.5.15),4K ProRes 422,Temporal NR - Better 2 Frames,7.60,FPS
Puget Systems DaVinci Resolve Studio Benchmark V0.6 BETA (DR 16.2.5.15),4K ProRes 422,3x Temporal NR - Better 2 Frames,2.64,FPS
Puget Systems DaVinci Resolve Studio Benchmark V0.6 BETA (DR 16.2.5.15),4K ProRes 422,Optimized Media,67.46,FPS
Puget Systems DaVinci Resolve Studio Benchmark V0.6 BETA (DR 16.2.5.15),4K ProRes 4444,Basic Grade,16.51,FPS
Puget Systems DaVinci Resolve Studio Benchmark V0.6 BETA (DR 16.2.5.15),4K ProRes 4444,OpenFX - Lens Flare + Tilt-Shift Blur + Sharpen,6.94,FPS
Puget Systems DaVinci Resolve Studio Benchmark V0.6 BETA (DR 16.2.5.15),4K ProRes 4444,Temporal NR - Better 2 Frames,7.56,FPS
Puget Systems DaVinci Resolve Studio Benchmark V0.6 BETA (DR 16.2.5.15),4K ProRes 4444,3x Temporal NR - Better 2 Frames,2.73,FPS
Puget Systems DaVinci Resolve Studio Benchmark V0.6 BETA (DR 16.2.5.15),4K ProRes 4444,Optimized Media,64.90,FPS
Puget Systems DaVinci Resolve Studio Benchmark V0.6 BETA (DR 16.2.5.15),4K RED,Basic Grade,12.33,FPS
Puget Systems DaVinci Resolve Studio Benchmark V0.6 BETA (DR 16.2.5.15),4K RED,OpenFX - Lens Flare + Tilt-Shift Blur + Sharpen,4.62,FPS
Puget Systems DaVinci Resolve Studio Benchmark V0.6 BETA (DR 16.2.5.15),4K RED,Temporal NR - Better 2 Frames,5.47,FPS
Puget Systems DaVinci Resolve Studio Benchmark V0.6 BETA (DR 16.2.5.15),4K RED,3x Temporal NR - Better 2 Frames,2.22,FPS
Puget Systems DaVinci Resolve Studio Benchmark V0.6 BETA (DR 16.2.5.15),4K RED,Optimized Media,26.06,FPS
```

Benchmarking Results

Interpreting Benchmarks

The Puget Systems benchmarks are based on a set of video files (included) shot in various resolutions using various codecs and may not match your camera's. While everyone seems to focus on the overall benchmark score, that's an average of the test results on five different codec types at various resolutions and frame rates (using certain effects). Poor performance with ProRes 444, CinemaRAW, or RED won't matter if you don't shoot in those formats. Ditto for 8K or 60 FPS. Likewise, slow processing of Temporal (Video) Noise Reduction won't affect you if you don't use that computationally intensive plug-in. Instead, pay attention to the codec your camera uses, the resolution and frame rate you shoot with, and the effects you are likely to use and ignore the rest. This means discounting the overall score and focusing on the few results that match your situation. Then and only then are the numbers meaningful. In fact, you can learn a lot from them.

When the Puget Systems benchmark says a certain effect can be processed at 45 FPS on your machine, that means the effect will probably playback on the fly on your computer at 30 FPS without having to halt and render first. If it's 22 FPS, for example, it'll play, but with glitches and stutters. If an effect processes at, say, 7 FPS, that means your computer won't be able to play it back without halting and rendering it out first.

Real-Time Playback

A better approach would be to create your own "benchmark" by shooting some sample footage with your camera and codec and monitoring the CPU, GPU, memory, and other hardware while performing the various types of post-production work you normally do and see if your setup can play the resulting Timeline at the correct FPS in real-time. That's the only accurate way of seeing how your camera's footage will actually perform on your PC's hardware. You can also use this method to get an idea of what you'll need if and when you decide to upgrade various components. I have a "benchmark" project that I use to test various settings changes and hardware tweaks to see if they really work.

In addition to running benchmark tests, it's a good idea to check your computer's performance with hardware performance monitoring software. Both MOBO and GPU card manufacturers have free software you can run and watch how your CPU, memory, drives, and GPU perform. It's good to know how often you hit 100% utilization on the GPU or CPU to know exactly where the hardware bottlenecks are. (However, do not leave these programs running in background while editing because they consume clock cycles.)

When I was looking for a new camera, Sony had four models that were essentially the same camera (same sensor, same glass, same features). But each one used a different codec. Which codec would work best on my hardware, given how I use Resolve? The only way to know that for certain was to download sample footage from each camera and test it. Thankfully, samples had been posted by somebody somewhere (Sony, are you listening?), and I was able to see exactly how each codec would perform on my setup.

If you need to build a PC specifically to handle Resolve, the go-to source for configuring PCs for Resolve is BMD itself. They have a very detailed configuration document for Linux, Windows, and Mac OS machines. You can get a copy from BMD's website. I would say that BMD's hardware recommendations are biased toward the high end. There are also several YouTube channels featuring Resolve that have detailed videos on hardware requirements and benchmarking test results.

For professional users, I would recommend buying a pre-configured machine from Puget Systems (www.pugetsystems.com). I have no relationship with them (financial or otherwise), but I have looked at their extensive information and advice (available for free online), and it has proven to be spot on.

Their thing is building PCs optimized for video editing (or other specific tasks). I took their advice (for free) and built my own, but realized well into the project that BIOS versions, graphics drivers, and various settings and tweaks were needed to get the best performance from the hardware—there's some black art at work here. In retrospect, what they do is very hard to duplicate on your own (it's not simply a matter of plugging cards in). If I had to do it all over again, I'd just buy a complete tricked-out, burned-in system from Puget and be done with it. As it happened, it took about two months of tweaking before my DIY machine was performing as expected.

Regardless of the benchmarks, the advice from various YouTube videos, or your local computer geek's thoughts, it is my considered opinion that if you are looking to upgrade your hardware, at minimum, you should aim for a computer capable of editing H.264/265 30 FPS HD media while achieving smooth playback and minimal halts to render and cache (unless you've applied a computationally intensive effect). If your computer can do that, you can edit 4K or higher res footage with transcoding, proxies, and other techniques.

While certain Resolve processes (noise reduction, grading, effects) can stress your machine more than others, the main performance constraints are the footage's codec, resolution, and frame rate. You are probably shooting at the highest resolution your camera can attain to achieve the best quality. And you've undoubtedly chosen a frame rate out of convention or necessity. So the only option you are going to be able to affect is the codec selection.

Tuning/Drivers

It's extremely important that you have the latest drivers for your MOBO, graphics card, and your OS is the most up-to-date version. Check with your computer's instructions, but most automatically update various drivers and other software if left on overnight.

In the case of Windows 10 machines, have a local computer geek tune your PC so that the CPU, memory, GPU, fans, and all the rest are running at top performance. Running Windows 10 Pro is preferred over the consumer versions because those are often chocked full of bloatware and unwanted apps that steal disk space and even clock cycles. Run performance monitoring software and make sure the install is clean and no other apps are running on the machine when you are using Resolve.

Overclocking should also be enabled. Usually, the periods of maximum performance (heat generation) while video editing are much less than for gaming. And the fan curve should be aggressive. When run wide open, CPU/GPU chips generate lots of heat. Onboard software will start to slow them down in order to keep them from getting too hot. Setting the fans to run sooner and faster will improve overall performance.

OS/App drive

The OS is used a lot by any machine, so having the OS on an SSD rather than an HDD is preferred for the speed advantage it gives. The Resolve app should be installed on this drive as well. A SATA 3.0 SDD is recommended. This type of drive loads a little faster than a standard HDD.

Project Drive

An NVMe drive for projects is needed because of the speed it can read and write your media files. You could put source media on the cache drive but having the project files, and the cache on separate NVMe SSDs can improve performance, particularly with large projects. For project media, I'm using a Samsung 970 EVO 500 GB NVMe SSD (in a PCIe M2 slot).

A SATA 3.0 SSD would work here but would still be limited to SATA speeds. That's faster than a SATA HDD but not as fast as an NVMe SSD for nearly the same price.

Roughly speaking, an NVMe SSD is about six times faster than a SATA 3.0 SSD, which is about five times as fast as an HDD. This means that an NVMe PCIe SSD can be as much as *35 times faster* than a typical HDD. Actual speeds depend on the motherboard's bandwidth, the number of PCIe lanes it uses, the particular slot it's installed in, and such like. At this writing, an NVMe SSD in a PCIe 3.0 M2 slot is about as fast a storage device as you can find for a consumer PC, but the cost premium as you ramp up in speed is not that great ($150 for a SATA SSD vs. $200 for an NVMe SSD). Of course, that's 4X the cost of an HDD, but the extra $150 is worth the 35X performance increase.

Cache drive

Resolve makes use of several types of intermediate render caches to improve playback performance. As noted, render files can be huge, and lots of them are generated during the edit. The cache needs to be on the fastest, largest drive you have. I'm using a Samsung 970 EVO 1 TB NVMe SSD (also in a PCIe M2 slot) for render caching. Resolve intermediate files are massive and can fill up a 1 TB drive quicker than you can say "wha?" Resolve can do some clean-up during and after the edit, but as of this writing, you'll have to police the cache drive yourself (particularly to remove Optimized Media and Proxy Media files). Delete the render cache when you are done with the project, and take a look at your cache drive from time to time to make sure it hasn't filled up with errant files.

Archive drive

For longer-term storage and backing up, an HDD drive is probably good enough. I'm using WD 2 TB HDDs for that. Resolve has an excellent system for archiving (covered later), but make sure you use Resolve's built-in archiving system instead of trying to save individual parts of a project via your OS's file management system.

CPU

A fast CPU with lots of cores is helpful for running any video editing system. I'm using a Ryzen 3rd Gen CPU (Ryzen 5 3600 6-core, 12-thread). It's AMD's entry-level 3rd Gen Ryzen CPU. I wish it had more cores, but given how I use Resolve, it's fine—for now. Despite the fact that most video processing is done on the GPU, Resolve does benefit from a faster CPU with more cores. If you are shopping for a new CPU, 8 cores seems to be the sweet spot, so a mid-level 3rd gen Ryzen would have been better. On the Mac, the M1 series CPU is getting great reviews from Resolve users. For Intel/AMD chips, John's Films on YouTube recently posted a video where he precisely determines the optimal number of cores for various Resolve 18 workflows.

GPU/Graphics Cards

Unlike other NLEs, Resolve makes extensive use of the GPU on the graphics card to process renders–it's not just for the video display. A few less-complex functions like transcoding (Optimized Media) are still done on the CPU or some combo thereof. Even so, a fast GPU and lots of video ram are essential if your projects are effects heavy because a so much of the video processing in Resolve is still done on the GPU. In fact, the performance of Resolve 15 through 17 was essentially determined by the GPU's performance. Beginning with Resolve 18, some of that load has been shifted back to the CPU. Ordinarily, that would be a bad move, but BMD did a major rewrite of the render engine in version 18, and it runs much better and faster on off-the-shelf hardware than previous versions.

For the graphics card, I'm using a Sapphire 5500XT. This is an AMD 3rd Gen card, but it's also their entry-level version, so it's not a powerhouse. But it's not expensive either.

With one GPU, my Puget Systems benchmark overall performance number was about 432 compared to 800+ for a fully tricked-out system built by John Daughtridge of John's Films for a feature film editor.

I added a second Sapphire 5500XT since the GPU was often running at 100% while the CPU and system memory were just above idle most of the time. But the overall score didn't change much. That's because performance on the ProRes 444, RED, and RAW files actually went down (AMD driver issues, I suspect). But performance for the file types I normally use (H.264 and ProRes 422) improved, making a nearly a 2X difference. With dual GPUs, almost all the effects I use now play back in real-time—which, ultimately, is the goal.

Whatever GPU you choose, make sure it does H.264/265/HEVC *hardware* encoding/decoding (assuming you need that), and make sure you tell Resolve it does. Note that hardware encoding/decoding not only happens when this media is first ingested, but almost every time those filetypes are accessed (unless you changed the codec via Optimized Media, Proxy Media, or Media Manager). In other words, whenever you are offered the option of hardware decoding, or encoding, take it.

Memory

I have 32 GB of 3200 CL16 RAM. You may need more than that, but start out with 16 or 32 GB and monitor your system's performance.

Motherboard/Chipset

All the above components run on a Gigabyte 570X motherboard (PCIe 4). This hardware choice is an attempt to future-proof the box when I upgrade hardware.

Whether your PC is brand new or not, make sure your BIOS and drivers are up-to-date and stay current. I've seen tremendous performance improvements as Gigabyte, Blackmagic, Microsoft, and AMD have tweaked drivers and application software since the release of the new AMD CPU/GPU lines. Of course, sometimes updates can cause a decrease in performance (until the update has been "fixed").

Monitors

Using dual monitors really helps because Resolve's user interface doesn't offer a lot of the re-sizing and workspace customization options you may be used to in Premier Pro or Final Cut Pro. So the more square footage you have in front of your eyes, the better. You can also run a "live video" full-screen monitor if you are using the Studio version. (As of this writing, the free version will also let you do that, but only if you have a DeckLink card from Blackmagic Design.)

Cooling

A large case with lots of fans is preferred because long renders, especially final, can really heat up the CPU and GPU if there's not enough air. So I put the above components in a Corsair Air box. As built, the chips run pretty cool, which is pretty cool. (The box is a beast, however.) I adjusted the fan parameters on the MOBO, CPU, and GPU so that the fans kick in early and fast because as the chips heat up, performance drops off.

Last Word

Many people (including me) do just fine with Resolve on a pre-M1 MacBook. With Resolve 18, you will have far fewer playback, rendering, or other performance issues. But Resolve, even version 18, does benefit from above-average hardware. A PC set up for gaming would be an ideal platform for running Resolve. To a very large degree, the hardware you need to run Resolve depends on the resolution, frame rate, and codec your camera uses, what you intend to do (basic editing effects vs. gee-whiz), and how you use Resolve (setup/workflow)—particularly the use of Timeline Proxy Resolution, various resolution and render settings as well as transcoding.

At the end of the day, properly setting up Resolve, performance tweaking your workflow, and managing renders (what, when, and how often) will almost certainly give better performance than simply throwing more hardware (and money) at the problem. If your PC plays the Timeline in fits and starts, take advantage of the many performance-enhancing tricks that are covered in the chapter on Resolving Performance Issues.

Whatever direction you plan to go in hardware-wise, run benchmarking tests and performance monitoring apps to fully understand where the actual bottlenecks are before doing any upgrades, and run benchmarking tests again after any hardware changes or software updates to make sure the upgrades actually improve things.

Resolving Performance Issues

Resolving Performance Issues

Being able to playback the Timeline (with various effects added) without stuttering and glitching or without having to halt and render can be an issue with any digital video editing system. In the past, DaVinci Resolve was especially problematic because BMD designs its products for professionals who are using high-end (and high-dollar) workstations and not necessarily regular folks with run-of-the-mill PCs. In fact, there were major sections in previous editions about getting DaVinci Resolve to run smoothly on average PCs.

With DaVinci Resolve 18, the render engine software has been completely rewritten, and version 18 runs much, much better on less-than-stellar hardware. Of course, you can add enough effects of enough complexity to bog it down, but generally speaking, Resolve 18 should run on a fairly recent PC or Mac without your needing to get under the hood and tinker. This chapter is included in case you need it. Again, you will experience stutters and glitches during playback and short breaks while it renders what you've just done. But if those are too frequent or too long, here's information that can help.

The Pause That Refreshes

Back in the day, when faced with a long render, I'd wait and schedule it for lunchtime, then punch the button and scare up a couple of colleagues to join me at a local eatery. Hopefully, when I returned, the machine would have done its thing, and the render will have finished. For shorter renders, I headed for the coffee machine.

We often talk about rendering as if it's an event instead of a continual process because when Resolve renders during playback, we don't notice it. It's like it's not even happening. The glitches and short pauses when Resolve is struggling to render and play at the same time, we do notice. And the long pauses, while Resolve stops playback completely while it renders, can be a real-time waster.

Getting the Timeline to playback smoothly on your PC (or Mac), without glitching or frequent work stoppages to render and cache, can be a challenge for new users. In fact, it's *the number one complaint* on Reddit. It's common for new users to make an edit or apply an effect, wait for the change to be rendered, play the clip, adjust the effect, wait for another render, and so on *ad infinitum*. But those stop-starts affect your workflow and disrupt your thinking process (at least it does mine). In an ideal world, everything would render during playback. But you can always add effects to a clip that no real-world PC can render and playback at the same time.

You can monitor real-time playback performance by looking at the Frames Per Second (FPS) indicator (top of the Viewer) while the Timeline plays. If your computer can handle the changes (rendering during playback), the indicator (really a GPU monitor) will be green, and the actual FPS will match your Timeline's actual frame rate. If not, the indicator will be red, and the FPS will be less than the Timeline's setting. You'll certainly notice the glitching and stuttering on the Viewer. In extreme cases, the Timeline may not play at all until the clip has been rendered and the results cached.

There are a lot of opinions about what computer you need for Resolve, and there are recommendations in the hardware chapter along with benchmark tests you can run. But at the end of the day, in order to use Resolve, your computer should be able to edit HD H.264/265 footage in an HD timeline without glitching too often and rarely halting for a render unless you've applied some electrifying effect that's computationally intensive. If your machine can do that, it will perform even better with a different codec (DNxHR, ProRes) or when operating at a reduced Timeline resolution or when using proxy files. The other performance issue is the behavior when Resolve needs/wants to render. That is determined by how the various render subsystems are set up (see previous chapter).

> **CAVEAT:** There are several options described below to deal with performance issues. If you read the previous chapters, you'll notice that much of the information here is reiterated but with greater detail, explanation, and amplification. However, if you are not having playback/rendering issues, you don't need this chapter at all (right now…).

You should have set preferences such that your computer will play back smoothly most of the time without constant pauses to render and cache. If the settings alone don't achieve that out of the box, this chapter will show you how to fine-tune Resolve so that it will render on the fly—during playback—as much and as often as your PC or Mac in its current state will allow.

Keep in mind that while more powerful hardware can help (and your hardware, particularly your graphics card, may, in fact, need upgrading), it's not impossible or even unusual to do things in Resolve that no real-world hardware can playback in real-time without ever having to halt and render. On the other hand, there are many preferences and settings in Resolve and workflow practices that can improve playback performance and reduce render times quite a bit, and you should always try these first before throwing more money at your PC.

Cache 22

When you are editing, grading, or adding effects, you are really only working with *one frame*—the one in the Viewer. Any PC built in the last decade can surely handle that. Any Mac laptop can, too. However, your computer's hardware is also expected to play the edited Timeline—which will certainly be more than one frame—and do so smoothly at 24, 30, or more frames per second.

The footage may have been shot on a high-resolution camera or at a high frame rate which will certainly challenge your machine. But you can also run into playback issues with less-expensive consumer or prosumer camcorders shooting in HD (1080p) because less-expensive cameras tend to use much more compression (to store more on an SD card) and with codecs that are easy for the camera to encode on the fly but are computationally more difficult to handle in post (such as H.264/265).

Digital video editing systems—every one of them—work with frames. But some in-camera codecs such as H.264/265 don't record actual individual frames, so the editing software must not only decode the compression scheme but also use that data to re-create individual frames so the Timeline can be edited or played.

It can take a fair amount of computational power just to play unedited footage straight from the camera. As you add transitions and effects, those are even more mathematically complex, requiring even more computations. Transitions and effects will have to be rendered to the equivalent of unedited camera footage, so your PC can again play the Timeline at the correct FPS.

Maths is the reason. A single HD frame is usually made from some 2 million individual pixels. With 8-bit video, each pixel will have one byte of information for each color (RGB) or 6 megabytes of data *per frame*. Your hardware will have to process something like 187 megabytes of information *per second* in order to render the clip playable in real-time. For 4K or 10-bit video, the numbers go up—way up.

It's not practical (or affordable) to build a PC capable of processing in real-time *anything* you might want to do to any video clip so that it will always play back at the correct FPS without halting and rendering it out first. That's why it's sort of a Catch 22—achieving smooth playback of complex effects can be a problem for both low-end consumer/prosumer cameras and high-end professional ones. And the more capable a video editing system is, the more things you'll want to do with it, and that means you'll need even more computing power to render out gee-whiz effects in high res in a reasonable amount of time.

While Resolve is an extremely powerful all-in-one video post-production system, all of this power comes at a price. The main drawback is that Resolve requires a fairly powerful computer (CPU), a very powerful graphics card or cards (GPU), a decent amount of memory, and, of course, very fast drives (SSDs) in order to play video at the correct FPS without dropping frames or stuttering or having to frequently halt and render. Throwing faster, more expensive hardware at the problem can improve playback and reduce the need to render as often but will never completely eliminate the need to occasionally halt and render (or cut render times to zero).

However, depending on several factors such as the complexity of the effect applied, various Resolve settings, and your graphics card's speed, most rendering can occur *during* playback, negating the need to halt and render (as often). You have no control over the effect's complexity—it is what it is. And while you can up your game graphics card-wise, for now, it is what it is, too. But you can set up Resolve and operate it so that it can render on the fly more often than not. Further, you can set up your workflow so that long renders take place when convenient to you and skip rendering (to a degree) when it's not.

The following guidance assumes you have done all you can do in the setup department to tune your computer for its maximum performance, and have not caused any problems yourself (i.e. scopes on while playing, other apps running in background).

First Things First

To improve Resolve's performance on your machine, the very first thing to do is get the paid version of Resolve (Studio). The free version is great (and it's free), but the free version does not handle H.264/265 files via your graphics card's hardware (if available). And H.264/265 files are hard to decode for editing. Resolve has to process H.264/265 files to make them playable (unless you've changed the codec using Optimized Media or Proxy Media). And not just when the files are imported. H.264/265 processing has to occur when an effect or transition is added or when the clip is playing. And even if you plan to transcode to a different codec, faster processing of H.264/265 files is a plus. So handling H.264/265 files using a hardware accelerator on a graphics card is a must if you shoot or edit with those codecs.

The paid version of Studio offers other advantages. Some functions in the free version are simply not implemented as well and do not run as fast as Studio. IMO, it makes little sense to upgrade hardware due to performance issues in order to try and get the free version of Resolve to work like Studio. In fact, the most effective performance upgrade you can do for Resolve is add a second GPU, and that can't be done with the free version.

What follows are recommended settings and operational practices to help Resolve render during playback (more often than not) on your machine:

Performance Mode

Performance Mode is set to Auto by default (Preferences > User >Playback Settings > Performance Mode). Basically, this app looks at your computer's capabilities and tries to optimize performance by reducing viewer resolution and adjusting other settings. It does not affect final output quality. It's also not that aggressive—you can make much more effective adjustments to various playback settings (below). Performance Mode also tells the rest of Resolve how much chutzpa your computer has, which affects when Resolve thinks a halt and render is necessary (in Smart Mode).

Monitor Settings

It's important to set your monitor's resolution to what it really is (or less), but certainly not higher (Project Settings > Master Settings > Video Monitoring > Video Format). If you are working in 4K, telling Resolve your monitor is only HD helps reduce the need to render as often and lets it perform fewer calculations when it does. Keep in mind that even if you have a 4K or 8K monitor, you are actually viewing the Timeline on a much smaller piece of real estate, so the image will not be 4K or 8K (unless you are full-screen).

Video bit depth should be 8 unless you really need to see 10-bit video in your viewer (and your monitor can actually display 10-bit video —most can't). Even if your camera records 10-bit video, you probably can edit with the monitor set to 8 bits if playback is an issue. Of course, editing on a higher res monitor makes everything look better (to you) but has no effect on the resolution of the final output and can unintentionally force Resolve into doing many more calculations than are necessary or even visible (which would mean more frequent, longer renders). Setting Video Monitoring to the full resolution your monitor can display usually only makes sense for full-screen viewing after you're done editing or for fine adjustments

> **CAVEAT**: Reducing the monitor's resolution is better than reducing the Timeline's resolution because the Timeline does not have to be re-rendered when you change the *monitor's resolution* back and forth. The monitor's resolution only affects what you see while editing—not the final product—and can be changed on the fly. If you reduce your monitor's resolution too much, the picture in your viewer may not look great, but the need to halt and render, and the render time when it does, will be less. If your eye offends you, you can always change it back.

Timeline Proxy Resolution

New with Resolve 18 is *Timeline Proxy Resolution* (Edit page *Playback*) which helps the Timeline play smoothly on the fly by reducing the resolution of the Timeline viewer (Edit page > Playback > *Timeline Proxy Resolution*). Timeline Proxy Resolution is like the Video Monitoring settings described above—but Timeline Proxy Resolution can be turned on and off from the Edit page with one click. Timeline Proxy Resolution is simply a Viewer resolution control and, despite the similarity in name, does not actually create Proxy Media. Timeline Proxy Resolution only affects the *Source and Timeline Viewers' resolution* and can be turned on and off at will. A slightly degraded (half resolution) viewer is usually fine for most editing, and you can turn Timeline Proxy Resolution off for full-screen viewing while doing detailed work.

I tested an unrendered clip that played at only 9 FPS with Timeline Proxy Resolution off. It instantly played at ~20 FPS with Timeline Proxy Resolution in half res and 22 FPS in quarter res. Quarter res is not pretty and isn't that much of a speed improvement, so I tend to set my Video Monitor to the Timeline's actual resolution (becomes the default), and I leave Timeline Proxy Resolution on in half res until I need to see a detailed view full-screen.

Timeline Resolution

If you are working in 4K or higher, you could set the Timeline resolution to 1080p or even 720p and work in a lower-res mode. Working in lower res helps the Timeline play at the correct FPS more often without the need to halt and render as much.

To do that, select the "gear" (lower right). Under Master Settings > Timeline Format, change the Timeline Resolution to something less than the original media. Also, set the monitor's resolution to match the Timeline's new resolution (or less). When you are done, change the Timeline resolution back to 4K (or whatever the original resolution is) before final output.

Working in a lower res Timeline and then outputting in full res can be an effective approach to dealing with playback issues. Yes, *the entire Timeline will have to be re-rendered* when you change the resolution back, but you can schedule that. IMO, that's more efficient than having lots of render interruptions while editing the piece. But test this set up with your own footage to see if the tradeoff of faster renders (but having to re-render the entire Timeline when done) is worth it to you.

> **IMPORTANT**: If you work on a high res project in a reduced Timeline resolution, *don't forget to change the Timeline resolution back before you output the project.* On the Deliver page, you can change the output resolution, but at that point, Resolve will simply upscale your project from the lower res Timeline. You don't want it to do that, so make it a part of your workflow to change the Timeline resolution back to the footage's original resolution before you hit the Deliver page.

Rationalize Resolution

There are several places In Resolve where various resolutions can be set: the Monitor, the Timeline, the Viewer, Render Cache Format, Optimized Media, Proxy Media, and Timeline Proxy Resolution.

For all these functions, there will be a rational set of resolution values that make sense and some that don't.

Setting the Monitor and Viewer bit depth or resolution *higher* than the Timeline's resolution, for example, doesn't make sense. Likewise, you wouldn't want to set the Timeline's resolution lower and set Render Cache to a compression setting designed for high res. You wouldn't want to edit 1080p footage in a 4K Timeline either.

On the other hand, setting the Monitor/Viewer resolution lower than the Timeline's resolution allows Resolve to still process the Timeline at the desired resolution while shortening the processing time of what you see on the Viewer during the edit.

I found that with 4K footage, on my computer, I could achieve real-time playback of most effects by setting the Timeline to HD, the Monitor and Viewer to HD, and Timeline Proxy Resolution to half—resulting in far fewer halt and render breaks. For delivery, I can either keep the Timeline in HD or change it back to 4K. Changing the Timeline's resolution does require a complete re-render, but you can work more or less uninterrupted in a lower res and let Resolve re-render the Timeline in full res when you are AFK or at the very end.

Proxy Media

Technically speaking, the term *proxies* refers to substitute files that have lower resolution than the original source and thus are easier to work with. During final output, proxies are used only as a guide and nothing more—the actual output file is built from the original source footage. It's similar to lowering the Timeline resolution, but you are doing it at the clip level where it affects every process going forward.

In Resolve, the system that creates actual proxy files is called *Proxy Media*. Typically, Proxy Media is used to convert high-resolution files to low res proxies that are easier on the hardware. But it can also transcode from H.264/265 to a codec your machine is better able to handle. The settings are the same as for Optimized Media (below), with the exception being that Proxy Media will allow you to create

H.264 files if you want. There must be a good reason for that option, but I can't think of one.

Optimized Media creates a proxy-like folder that can be shared on the same network, but Proxy Media creates standard .mov, AVI, or MP4 files that can be shared with another editor along with the rest of the project's files. That negates having to re-generate them if you have a collaborator.

Optimized Media

Like Proxy Media, Optimized Media can transcode the codec used by your camera to one that's easier for Resolve to work with. In that case, you could be working with a codec your camera does not actually have.

When you optimize camera footage, not only can you choose which codec to use, you can also generate optimized files at a lower resolution than the originals. That also improves playback and render performance.

Whether Optimized Media will improve your playback performance or not depends on the resolution, frame rate, and codec your camera uses and how fast your graphics card can process that native footage. Simply changing from a hard-to-edit codec (H.264/265) to one that's easier to work with (ProRes/DNxHR) can achieve dramatic improvements in playback performance. You also have the option of reducing the resolution of the footage and using higher compression settings.

I tested an XAVC S clip (H.264 long GOP in an MP4 wrapper) vs. AVCHD (H.264 in an MP2 wrapper). Optimized Media did help improve on-the-fly playback and render times when using AVCHD files. But I saw less improvement in render times using Optimized Media on XAVC S clips. The better your computer can play H.264/265 files, the less Optimized Media is going to help.

In the tests I ran, reducing both the monitor and Timeline resolution improved playback performance substantially, but using very high compression on Optimized Media files had little to no effect on playback. But the files will be smaller, and you will fill up your cache drive less often.

Transcoding via Optimized Media (or Proxy Media) takes time—from somewhat less than real-time up to a minute or longer per minute of run time. So the *length* of a clip is the determining factor of how long optimization will take. One way around that is to create subclips and optimize just the parts you intend to use. Another way is to drag a clip to the Timeline, and cut it down to just the segment you need. Right-click on the now shorter clip and *Generate Optimized Media* directly in the Timeline.

If you do intend to use Optimize Media, you could process all your files at once when it's convenient for you. It's usually better to have a long transcode at the beginning and a long render at the end when you can schedule it than the constant interruptions. Of course, your camera may allow you to create video files using a codec that places less workload on your graphics card. If so, try the various codecs your camera offers and see which one (if any) improves Resolve's performance.

Under the "gear" (bottom right), select Master Settings. Set the Optimized Media Resolution, the Optimized Media Format, and the Render Cache Format to match the resolution of the source footage (or lower to improve performance during the edit). Creating lower res files won't affect final output quality (unless you intentionally use them for final output). DNxHR(something) is the best choice for PC users while ProRes is the proper choice for Mac users. Here the "something" is the actual DNxHR compression setting. Lower is faster and higher is better quality (but, again, only you will see the degraded view on your editing viewer).

The only way to know for sure if Optimized Media is worth to you is to optimize a short clip shot on your camera (using Original and reduced resolution and various compression settings). Add some effects that are known to be troublesome (or ones you intend to use). Then try playing the Timeline while monitoring the FPS or rendering times, and see if *Optimized Media* indeed improves playback performance and which settings are optimal for your workflow.

If you do go this route, delete Optimized Media when you're done with the project. Optimized Media files are persistent and can take up massive amounts of space. You used to be able to delete Optimized Media when you Archived the project. Beginning with Resolve 17, this critical function was taken away. Resolve 18 didn't return it, either. So, for the time being, at least, you'll have to manually delete Optimized Media files using the standard file deletion tools in your OS. Navigate to your cache drive and from there to Davinci Resolve Cache > CacheClip > OptimizedMedia.

Most people are shocked at how big those files actually are. Despite best efforts, loose bits can pile up in your cache drive and slow its performance. If your machine starts balking, clean out the cache drive, and use your OS's disk optimization tools to get it back in spec.

In-Camera Proxies

Another option is to record proxies in the camera (if available), or external recorder, and link them to the full-resolution footage in Resolve. That saves lots of time. However, using camera-generated proxy files and linking them to the original source footage ain't for the faint of heart. But, then again, that may be the only practical solution until you can get a camera that records in an easier-to-edit codec or use an external recorder. The official Reference Manual describes in detail how to do it.

Scopes

The video scopes are handy and at times, even necessary, but they require lots of computations, and having the scopes on while playing the Timeline could overwhelm the capabilities of the graphics card to render on-the-fly with earlier versions of Resolve. I haven't found that to be true with Resolve 18, but if having the scopes on all the time seems to drag your computer down, turn off scopes during playback. Leave them active the rest of the time if you want (but you really only need them while making adjustments).

And it goes without saying that you should not have any other processes running in background on your PC when running Resolve (like e-mail or text messaging).

Show All Video Frames

On the Edit page, under the three dots at the top right, there is the *Show All Video Frames* control.

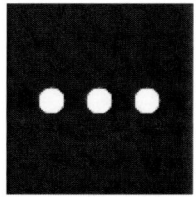

3 Dots

The default is *Show All Video Frames*. That prioritizes trying to playback every single frame of the video at the expense of choppy audio. If you uncheck it, Resolve will try to playback the *audio track* as smoothly as possible at the expense of glitchy video by skipping problematic frames. In fact, if your playback is already close to the correct FPS, simply unclicking *Show All Video Frames* may allow "good enough" on-the-fly playback. Of course, there'll be a dropped frame here and there, but in most cases, that's not a problem because only you see the glitches (the final render for Delivery will clean all that up).

Which mode you prefer is up to you. I'm guessing that at different times, one setting will make more sense than the other. That said, this control only affects Timeline playback on your machine, so flipping back and forth has no downstream consequences. Usually, I leave *Show All Video Frames* off—but not always.

Adopt a Workflow

Finally, it's important to adopt a workflow that accommodates your computer's capabilities (or lack thereof). This varies with the media you are working with as well as your computer's CPU, GPU, and memory, but in general, with Render Cache set to Smart, if there's too much halting and rendering going on, set it to User and render just the clips you need to by enabling or disabling Render Cache Color Output per clip. Or set it to None while you edit and back to Smart when it's convenient to let Resolve halt and render everything out.

I sometimes shoot with cameras that only offer H.264/265-based codecs. So my first order of business is to transcode via Optimized Media (or, more recently, Proxy Media). That can take quite a while if there's lots of footage, so I batch transcode and use the time to put up gear, recharge batteries, wash the van, get a haircut. I typically (but not always) edit in a reduced-resolution Timeline and playback performance is generally good. The final render can also take a while, so I schedule that when I can efficiently do something else productive.

At the end of the day, workflow is a critical performance factor. Either you'll have to adopt a workflow that accommodates your camera (resolution, frame rate, codec), or you'll have to upgrade to a ridiculously powerful machine. It's your checkbook and your choice.

Finally, when experimenting with a new effect, it's best to *use a very short clip*. I applied the new Intuitive Object Mask to a 15-minute clip, and the processing time was estimated to be nearly two hours. So I cut off a 10-second subclip and practiced with that. Once you learn Resolve, you'll know what effects can be added to a clip and the Timeline will still play on your computer without having to halt and render, and which can't. I frequently use snippets of clips to try various effects and processes and work out the settings before applying them to the entire clip. In some cases, I'll hold off applying computationally intensive effects until the very end or when I need to take a break.

Render Systems

RENDER SYSTEMS

Rendering

Have you ever seen a motion picture? No, you haven't. Pictures don't move. What happens is you are shown a series of *still pictures,* and your eye/brain "renders" that information to make you think the pictures are actually moving. Rendering by computer is sort of the same thing. A data file containing visual information is processed so that the result is a continuous series of images that are played back at some number of frames per second (usually 30).

To save disc space and for other reasons, the actual data file is not made up of 30 complete pictures per second. When you hit the play bar, your computer has to take that data and convert it—on the fly if possible—into 30 frames (pictures) every second. Most consumer and prosumer cameras compress and record video in such a way that rendering must be done every time the video is played. When editing, we often add transitions (dissolves), text, graphics, and effects that also have to be processed so the resulting video will play back at 30 frames per second (or whatever the frame rate is).

If your computer is not fast enough, some combination of compressed video plus added effects may not render as fast as the video is supposed to be playing. If it's close, the video may occasionally stutter. If the effects are computationally intensive, Resolve may want to stop playing until it can render the Timeline so that it will play smoothly at the correct frame rate. While some renders are mandatory (such as when Deliver creates the final output file), constant halting and rendering while you are trying to edit is frustrating and can make you lose your train of thought.

As has been mentioned, one of the biggest changes with Resolve 18 is the complete rewrite of the render engine (the software that does all that). Previous versions of Resolve required users who had less than state-of-the-art PCs to know quite a bit about Resolve's various rendering systems and also how to use them properly. For most people, that's no longer the case. You should expect the occasional stutter or glitch, and having Resolve stop and render a newly added effect is not unusual either. However, by turning various rendering systems on and off during the edit, by adding effects and rendering when it's more convenient, by editing in a reduced resolution, and taking other steps, you can reduce the number of times Resolve will have to stop everything and render the clip so that it can play smoothly.

The Timeline Render Cache

The render system you'll work with most often is Render Cache. BMD's documentation sometimes calls this the "sequence" cache. At other times, BMD calls it the "render" cache or the "Timeline render cache." Whatever you call it, it's the system that renders transitions and other changes to the Timeline performed on the Edit page (mostly). (Despite BMD's documentation stating otherwise, Render Cache also renders OpenFX and OFX effects if applied via the Edit page.)

Resolve's render behavior on the Edit page depends on the settings of both Render Cache, Render Cache Fusion Output, and Render Cache Color Output. The Optimized Media and Render Cache panel in Project Settings also plays a role.

You control Render Cache's behavior via Playback > *Render Cache*. There are three options: (1) None (off), (2) Smart, and (3) User. This control is basically the Resolve render system's "master switch" because it can affect render behavior of Timeline/Edit page rendering as well as the rendering of effects added via the Fusion and Color pages. It's this behavior that can be confusing. Render Cache is indeed the render controller for transitions and effects added via the Edit page, but these master settings (Smart, User, None) also enable or disable rendering for pages downstream (Fusion, Color) when their rendering switches are set to Auto.

Render Off

If you set Render Cache to *None*, Timeline caching will be turned off, and Resolve will not automatically halt and render regardless of other render settings. A red line will not appear over a clip that needs rendering either. With Render Cache off, you are forcing Resolve to only render on the fly, in real-time, during playback—*if it can*. If your computer is powerful enough, you may not need to halt and render, although playback will probably stutter occasionally. If your computer can't play the Timeline smoothly to your satisfaction, you'll need some implementation of Render Cache on (Smart or User).

Smart Mode (Automatic)

With Render Cache in Smart mode, a red progress line will appear over any clip that *Resolve thinks* needs to be rendered before it will play smoothly on your machine. During playback, the red line will change to blue as the clip is rendered. When it's all blue, it will play at the Timeline's correct FPS without glitching or stuttering (mostly). In Smart mode, Render Cache can also change the progress line from red to blue when you are AFK (not moving the mouse).

Occasionally, certain types of media will have to be rendered when first pulled into the Timeline even though you've done nothing to them yet (that's really a job for Optimized Media). On my machine, .mov files from my smartphone need to be rendered as soon as they're dragged to the Timeline.

If you selected Smart Render and a red line does not appear over a clip, Resolve's AI believes it will play smoothly on your machine in real-time. If Smart mode doesn't work the way you think it should, you can manually force a render by right-clicking on a clip and enabling *Render Cache Color Output* (this is known in Resolve World as "flagging" the clip. Occasionally, you'll still have to switch to User mode and flag the clip (anomalous behavior).

User Mode (Manual)

If you set Render Cache to *User*, a red progress line will not appear above a clip unless you've also enabled Render Cache Color Output for that clip (i.e., you have "flagged" the clip). In that case, a red line will appear and rendering will proceed.

This mode is preferred if your machine has enough power to normally play Edit page operations and effects without glitching too badly or too often. When it can't, you can always right-click on the clip and select *Render Cache Color Output* to force a render.

You can allow transitions, composites, and Fusion effects to be automatically cached even in User mode (or disable this behavior). The controls are in Project Settings > Master Settings > Optimized Media and Render Cache.

Because there are several render/cache systems in Resolve, and they are interconnected, here's a table to help you sort them all out.

Render Cache (Timeline/Edit Page)

Render Cache Smart Mode

Clips that need to be rendered (because of effects that were added via the Edit page) are automatically identified (red line) and rendered/cached (blue line) during playback (if feasible) or while you are AFK if "background" caching is enabled in Project Settings.

Exceptions: Resolve may wrongly think a clip can play smoothly on your computer when it actually doesn't. Likewise, Resolve may want to halt and render a clip that, in reality, plays just fine or with a few stumbles. If that creates a problem, set Render Cache to User and disable/enable *Render Cache Color Output* for each clip as needed.

Render Cache User Mode

Renders during playback or AFK if the clip's *Render Cache Color Output* is enabled.

Render Cache Fusion Output

Render Cache in Smart Mode
When Render Cache Fusion Output is set to Auto, and Render Cache is in Smart Mode, Resolve automatically caches H.264/265, HEVC, and certain raw camera formats, speed and retime effects, titles and generators, and Fusion effects applied via the Fusion page (basically anything applied upstream of the Color page). You can manually force a Fusion render cache by enabling *Render Cache Fusion Output* by right-clicking on a clip and turning it on..

Render Cache in User Mode
Caching is (generally) not automatic. Renders any clip that has Render Cache Fusion Output manually enabled.

Exception: Even in User mode, Resolve will cache transitions, composites, and Fusion effects automatically if enabled in Project Settings > Master Settings > Optimized Media and Render Cache. (*Automatically Cache Fusion Effects in User Mode* is On by default.) See BMD's Reference Manual for further details.

Fusion also has a Node caching system, but it's beyond the scope of this book. The DaVinci Resolve 18 Reference Manual has lots of info about this if you are curious. But it's very similar to Node caching on the Color page.

Render Cache Color Output

Render Cache in Smart Mode
Renders Motion Blur, Noise Reduction, OpenFX, or OFX effects (even if applied via the Edit page) when a Node's *Node Cache* is set to Auto (the default) or On. You can manually "flag" a clip for rendering by right-clicking on it and enabling *Render Cache Color Output*.

Exception: You can enable or suppress rendering of a specific Node by right-clicking on a Node and turning Node Cache to On or Off. When a Node Cache is set to Auto, the higher level Render Cache settings apply.

Render Cache in User Mode

Caching is not automatic. You must manually "flag" a clip for rendering by right-clicking on it and enabling *Render Cache Color Output*. Also Renders Motion Blur, Noise Reduction, OpenFX, or OFX effects (even if applied via the Edit page) when a Node's *Node Cache* is set to On as well as any Nodes upstream of the Node being cached.

Exception: You can enable or suppress rendering of a specific Node by right-clicking on a Node and turning Node Cache to On or Off. When a Node Cache is set to Auto, the Render Cache settings apply.

Node Caching

All of the above caching systems and controls operate at the clip level. The Fusion and Color pages give you a third option—caching at the *Node level*. Some Nodes contain more computationally intensive effects than others, and when caching the Color page's output, those Nodes are cached again and again. Often, once is enough.

It's a common practice to put computationally intensive effects (such as Video Noise Reduction) on the first Node and cache it. Then, when downstream Nodes are added and also require caching, at least the hard one(s) have already been cached.

To cache a Node, right-click on it and select On. The label will turn red. It turns blue when it's cached. The Render Cache high-level control determines when any caching takes place, so Render Cache needs to be set to Smart or User. Typically, I create a series of Chroma Key Nodes and cache the last one, and a series of grade Nodes and cache the last one of those. Then I can add several "clean up" Nodes and adjust those without having to re-cache the earlier (harder) ones.

Basically, when you cache a Node, you are caching that Node and all the ones upstream of it. Node caching can save lots of time, but if you have to change something, the Node will have to be re-cached. For this reason, it's a good practice to put static Nodes in early and cache them. Unless the input to a Node changes, it won't have to be re-cached when the Timeline renders.

Best Practice

My advice for new users is to set the Render Cache to Smart. With a reasonably powerful machine, most rendering will take place on the fly anyway, particularly with Resolve 18. Smart rendering usually handles the rest. When necessary, you can force a render by setting Render Cache to User and enabling Render Cache Color Output (right-click) for a specific clip.

The second option is to set Render Cache to User and cache on a clip-by-clip basis as necessary (right-click on a clip and "flag" it for rendering by enabling Render Cache Color Output). Alternatively, you could temporarily suspend background rendering altogether while you complete your opus, then turn it back on and see the results (eventually).

When you finish a project, delete the Render Cache (Playback menu), or your cache drive will quickly fill up (the folder is DaVinci Resolve Cache if you want to check it out). Deleting Render Cache from the Playback menu is *per project*, so deleting "All" only deletes them for the open project. It's still a good idea to check your Render Cache folder from time to time and clean out any extraneous junk.

Render in Place

Mentioned previously is Render in Place. Sometimes, clips get rendered over and over again, even if no changes have been made. One way to fix that is to Render in Place. That creates a rendered version of the clip that doesn't have to be re-rendered unless you change something. If you have added a computationally intensive effect (you'll know when you do that), you can that clip using Render in Place, and it won't have to be re-rendered the next time the Timeline gets rendered.

Last Word

I apologize for the long and winding road, but Resolve, for all its good qualities, does not always put "like with like" and sometimes hides essential settings in places you'd never think of looking (and names things the opposite of what they really are). Resolve also uses idiosyncratic and inconsistent naming conventions.

Made in United States
North Haven, CT
12 February 2023